T0143394

High Temperature Electronics Design for Aero Engine Controls and Health Monitoring

RIVER PUBLISHERS SERIES IN CIRCUITS AND SYSTEMS

The "River Publishers Series in Circuits & Systems" is a series of comprehensive academic and professional books which focus on theory and applications of Circuit and Systems. This includes analog and digital integrated circuits, memory technologies, system-on-chip and processor design. The series also includes books on electronic design automation and design methodology, as well as computer aided design tools.

Books published in the series include research monographs, edited volumes, handbooks and textbooks. The books provide professionals, researchers, educators, and advanced students in the field with an invaluable insight into the latest research and developments.

Topics covered in the series include, but are by no means restricted to the following:

- Analog Integrated Circuits
- Digital Integrated Circuits
- Data Converters
- Processor Architecures
- System-on-Chip
- Memory Design
- Electronic Design Automation

For a list of other books in this series, visit www.riverpublishers.com

High Temperature Electronics Design for Aero Engine Controls and Health Monitoring

Lucian Stoica
GE Global Research
Germany

Steve Riches
Tribus-D Ltd
UK

Colin Johnston
University of Oxford
UK

Published, sold and distributed by:
River Publishers
Alsbjergvej 10
9260 Gistrup
Denmark

River Publishers
Lange Geer 44
2611 PW Delft
The Netherlands

Tel.: +45369953197
www.riverpublishers.com

ISBN: 978-87-93379-25-1 (Hardback)
978-87-93379-24-4 (Ebook)

Contents

Preface xi

Acknowledgments xiii

List of Figures xv

List of Tables xxi

List of Abbreviations xxiii

PART I: High Temperature Electronics Background
Lucian Stoica, Steve Riches, Colin Johnston

1 High Temperature Electronics for Aviation Applications 3
1.1 Value Story 3
1.2 Fuel Prices Are Challenging the Airliners Profitability 4
1.3 Growing Fuel Efficiency 4
1.4 Clean Sky Initiative 6
1.4.1 Benefits of High Temperature Electronics for Jet Engine Controls and Health Monitoring 7

2 High Temperature Integrated Technologies 11
2.1 Introduction 11

PART II: Development of Multi-Sensor Data Acquisition System
Lucian Stoica, Valentyn Solomko, Ozan Iskilibli, Renato Del Regno, Reece Beigh, Thorsten Baumheinrich, Steve Riches, Colin Johnston, Geoff Rickard, Paul Williams

3 Outline of System 19
3.1 High Level Input Specification 19
3.2 Technology Assessment and Selection 19

3.3 Definition of Prototype System 20
3.3.1 HIGHTECS SOI ASIC 21
3.3.2 HIGHTECS Hybrid Circuit 23
3.3.3 HIGHTECS Module 24
3.4 Manufacture of Prototypes 24
3.4.1 HIGHTECS ASIC in PGA Package 24
3.4.1.1 HIGHTECS hybrid circuit 24
3.4.1.2 Assembly of Ceramic Substrate to Metal Package 27
3.4.1.3 High Temperature PCB for Resistors . . . 29
3.4.1.4 HIGHTECS Module 29

4 Design and Characterization of HIGHTECS Signal Channels and Building Blocks 31
4.1 Operational Amplifiers 31
4.1.1 Rail-to-Rail OpAmp 32
4.1.1.1 Schematic diagram 32
4.1.1.2 Layout 33
4.1.1.3 Simulation results 34
4.1.2 PMOS-input OpAmp 34
4.1.3 NMOS-input OpAmp 36
4.2 Bandgap Reference Generator 38
4.3 Bandgap Voltage and Reference Current 40
4.4 Bias Network 41
4.4.1 Top Level Schematic Diagram 41
4.4.2 Bias Network Layout 44
4.4.3 Reference Voltages Generator 45
4.4.4 Voltage to Current Converter 46
4.4.5 Current Mirrors 46
4.5 Analog Multiplexer 48
4.5.1 Layout of Analog Multiplexer 50
4.6 Single-Ended to Differential Converter 50
4.6.1 Simulation Results 50
4.6.2 Layout of Single-Ended to Differential Converter . . 53
4.6.3 Single-Ended to Differential Converter 54
4.6.4 Measurement Results 54
4.7 T1/TFo — Temperature Channels 57
4.7.1 Temperature Channels 57
4.7.2 T1/TFo Operating Principle 59

4.7.3 T1/TFo Functionality 59
4.7.3.1 Voltage-gain profile 59
4.7.3.2 Static accuracy 60
4.7.3.2.1 *Static accuracy simulation for variable chip temperature and constant sensor temperature* . . 61
4.7.3.2.2 *Static accuracy simulation for variable chip temperature and variable sensor temperature* . . 62
4.7.3.3 Temperature error due to quantization error 62
4.7.3.4 Input ESD protection 62
4.7.3.5 Short and open circuit detection 63
4.7.4 Mirrored Bias Current for Temperature Probe Excitation 63
4.7.5 T1/TFo Channels Schematic Diagrams 64
4.7.5.1 T1 top level connection 64
4.7.6 T1/TFo Channels Layout 67
4.8 SG2 — Strain Gauge Channel 68
4.8.1 Testing of HIGHTECS Module 68
4.9 QFREQ — Frequency Channel 69
4.9.1 Introduction 69
4.9.2 System Architecture 71
4.9.2.1 Input signal definition 73
4.9.2.2 Theoretical performance 74
4.9.3 Circuit Design and Implementation 75
4.9.3.1 Input circuit 75
4.9.3.2 Current detect path 80
4.9.3.3 Voltage path 82
4.9.3.4 Pulse selector 84
4.9.4 Experimental Results 85
4.9.4.1 High voltage amplifier 90
4.9.4.2 Measurement results 90

5 Characterization of Prototypes 95
5.1 Assessment of Prototype Performance 95
5.1.1 ASIC in PGA Package 95
5.1.2 Functional Tests 95
5.1.3 Design Changes Implemented for 2nd Version of HIGHTECS ASIC 96

5.1.4 Modification to Top Layer Metallisations on 1st Version of HIGHTECS ASIC 97
5.1.5 2nd Version of HIGHTECS ASIC 97
5.1.6 Environmental Tests 98
5.1.7 Characterisation Tests 98
5.1.8 Prototype SiC Transient Voltage Suppressors 98
5.2 Testing of SOI Test Chip 100
5.2.1 High Temperature Storage (250°C) 102
5.2.2 Rapid Temperature Cycling (–40°C to +225°C) . . . 103
5.2.2.1 Profile No. 1: 4 × following cycle 104
5.2.2.2 Profile No. 2: 2 × following cycle 104
5.2.3 Vibration (Room Temperature and 200°C) 107
5.2.4 Silicon Capacitors 107
5.3 Functional Tests on Eagle Test Systems 108
5.3.1 Room Temperature Testing 108
5.3.2 High and Low Temperature Testing 110
5.3.3 Environmental Tests 110
5.3.3.1 High temperature storage (200°C and 250°C) 110
5.3.3.2 Temperature cycling (–40°C to +250°C) 111
5.3.3.3 Vibration/Shock 112
5.3.3.4 Testing of HIGHTECS hybrid circuit and high temperature PCB containing resistors 113

6 Reliability, Failure Rates and Lifetime Prediction **119**
6.1 Accelerated Life Tests and Lifetime Prediction 119
6.1.1 Thermal Ageing at 200°C and 250°C 119
6.2 FMEA and Reliability Prediction 119
6.2.1 Module Weight and Dimensions 119
6.2.2 Module Power Consumption 121
6.3 Summary 121

7 Future Directions for High Temperature Electronics **123**
7.1 Semiconductor Devices 123
7.2 Passive Components 123
7.3 1st and 2nd Level Assembly 124
7.4 Custom Metallisations 124

7.5 EMI/Lightning Protection 125
7.6 Applications 125
7.7 Commercial/Environmental Factors 126
7.7.1 Market Size 126
7.7.2 Custom vs Discrete Solutions 126
7.7.3 Integration into Systems 126
7.7.4 Lifetime Support 127
7.7.5 Economics 127

Index **129**

About the Authors **131**

About the Contributors **133**

Preface

Incorporating electronics into hotter parts of equipment for monitoring and measuring outputs from sensors in aircraft engines or for down-well drilling and logging has been the goal of several organisations worldwide, with the objectives of reducing the amount of cables and eliminating the need for cooling systems, leading to increased fuel efficiency and reduced gaseous emissions. The challenges include finding semiconductors that can work at temperatures of 200°C and above, the availability of passive components and having reliable packaging and interconnections that can withstand the high temperature environment for the lifetime of the product (up to 25 years for aircraft).

This book is the culmination of work carried out within an EU Clean Sky project called HIGHTECS, which realised an Application Specific Integrated Circuit (ASIC) to carry out the signal conditioning and processing from a range of sensors representative of an aero-engine application and fabricated using a Silicon-on-Insulator (SOI) semiconductor process. The ASIC functionality was characterised over a range of temperatures from –40°C to +250°C and demonstrators were built using high temperature electronics assembly techniques.

Finally, some thoughts on the future applications of high temperature electronics and the issues influencing more widespread commercial exploitation are presented.

Acknowledgments

The results presented in this book were generated during High Temperature Survival Electronic Devices for Engine Control Systems (HIGHTECS) project.

HIGHTECS was partially funded by EU Clean Sky Grant 255749 and support from GE. The specification was defined by Turbomeca and the project was carried out by a consortium of GE-Research Munich (Germany), GE Aviation Systems (Newmarket, U.K.) and Oxford University (U.K.).

Disclaimer

Book Abstract

This book will cover the development of a demonstrator distributed high temperature electronics platform for integration with sensor elements to provide digital outputs that can be used by the FADEC (Full Authority Digital Electronic Control) system or the EHMS (Engine Health Monitoring System) on an aircraft engine.

List of Figures

Figure 1.1 Aero engine fuel saving scheme based on [5, 6]. . . 5
Figure 3.1 Block diagram for SOI ASIC in HIGHTECS module. 21
Figure 3.2 1st Version of HIGHTECS ASIC – device size 7.48 mm × 5.95 mm. 22
Figure 3.3 Layout of HIGHTECS hybrid circuit. 23
Figure 3.4 Mechanical assembly drawing for HIGHTECS module. 25
Figure 3.5 Silicon wafer containing HIGHTECS ASICs. . . . 26
Figure 3.6 HIGHTECS ASIC assembled in HTCC PGA package. 26
Figure 3.7 HIGHTECS hybrid circuit substrate. 27
Figure 3.8 HIGHTECS populated hybrid circuit substrate. . . 28
Figure 3.9 HIGHTECS hybrid circuit mounted in metal package. 28
Figure 3.10 Resistors surface mounted onto high temperature printed circuit board. 29
Figure 3.11 Stainless Steel enclosure with mounted PCB and hybrid circuit. 30
Figure 3.12 Stainless steel enclosure with lid incorporating EMI gasket. 30
Figure 4.1 Rail-to-Rail Class-AB Output Stage Opamp Schematic. 32
Figure 4.2 Rail-to-Rail Class-AB Output Stage Opamp Layout. 33
Figure 4.3 Schematic of the PMOS input opamp with class-AB output stage. 35
Figure 4.4 Circuit schematic, device sizes and bias current of the NMOS input class AB output stage opamp. 37
Figure 4.5 Circuit schematic, device sizes and bias current of the 3 input opamp. 39

Figure 4.6 Symmetrically matched current-voltage mirror to generate V-reference. 39
Figure 4.7 Bandgap voltage (actual and percent change) vs. temperature. 40
Figure 4.8 Layout of the bandgap voltage reference cell. . . . 42
Figure 4.9 Post-layout extraction simulation results of the bandgap voltage cell over PVT corners. 43
Figure 4.10 Bias network layout. 44
Figure 4.11 Reference voltages generator schematic diagram. . 45
Figure 4.12 Schematic diagram of voltage to current converter. 45
Figure 4.13 Nominal simulations for small-signal stability of the voltage to current converter. 47
Figure 4.14 Analog multiplexer schematic diagram. 48
Figure 4.15 2:1 Multiplexer and transmission gate implementation. 48
Figure 4.16 Multiplexer simulation results. 49
Figure 4.17 Layout of analog 11:1 multiplexer. 51
Figure 4.18 Nominal simulation results for single-ended to differential converter. 52
Figure 4.19 Single-ended to differential converter layout. . . . 53
Figure 4.20 Schematic of the single-ended to differential converter. 54
Figure 4.21 Micrograph of the designed instrumentation amplifier and single-ended to differential converter in X-FAB XI10 SOI process. 55
Figure 4.22 Measured DC gain of the instrumentation amplifier used in the strain gauge channel. 55
Figure 4.23 Measured linearity of the temperature channel at 225°C. 56
Figure 4.24 Measured transfer function of the single-ended to differential converter at 225°C. 56
Figure 4.25 Measured output waveform of the strain gauge channel with a 16 mV sinusoidal input indicates a gain of 240. 57
Figure 4.26 Temperature channel T1 signal conditioning diagram. 58

Figure 4.27 Voltage measured across TFo2 terminals vs. temperature. 58
Figure 4.28 Voltage gain profile of the analog front-end. Blue area – extreme voltages, corresponding to 2.5 mA/ 3 mA excitation current. Green area – nominal profile, corresponding to 2.7 mA excitation current. 60
Figure 4.29 Output voltage of T1 channel. Transistor-level bandgap and excitation current source. The largest static error over three simulated cases is 4.5°C (±2.26°C) at –60 . . . +250°C temperature span, and 1.41°C (±0.7°C) at +50 . . . +150°C. 62
Figure 4.30 Output voltage of T1 channel. Ideal excitation current source and reference voltage source. The largest static error over three simulated cases is 1.76°C (±0.88°C) at –60 . . . +250°C. . . 63
Figure 4.31 Simulated resistance error. Transistor-level bandgap and excitation current source. The largest static error over three simulated cases is 0.67 Ohm, which corresponds to 1.77°C (±0.89°C) at –60 . . . +130°C. 65
Figure 4.32 Input pad for channel T1/TFo (“APRBDF”). 65
Figure 4.33 T1 top level schematic. 66
Figure 4.34 TFo2 $I_{excitation}$ (solid) and mirroring ratio (dotted) vs. temperature. 66
Figure 4.35 T1/TFo layout – size: 730 um × 1530 um. 67
Figure 4.36 Simplified schematic of the strain gauge signal conditioning channel. 68
Figure 4.37 HIGHTECS module. 69
Figure 4.38 Peak detector presented in Figure 9 of [9]. (c) Springer. Reprinted with permission. 70
Figure 4.39 Peak detector presented in Figure 2 of [11] (c) IEEE. Reprinted with permission. 71
Figure 4.40 HIGHTECS ASIC function level block diagram including the signal conditioning processing the high voltage frequency signal. 72
Figure 4.41 Block diagram of the frequency signal conditioning unit for rotating equipment. 73
Figure 4.42 Input signal model. 73

Figure 4.43 System averaged frequency error simulation results for a maximum input frequency F_{sig} = 3999 Hz and a reference clock frequency F_{ref} = 10 MHz. 76
Figure 4.44 System averaged frequency error simulation results with added uniformly distributed jitter between ±25 ns for a maximum input frequency F_{sig} = 3999 Hz and a reference clock frequency F_{ref} = 10 MHz. 77
Figure 4.45 Top level circuit schematic of the signal conditioning unit processing the high voltage frequency signal. 77
Figure 4.46 Input stage circuit schematic. Diodes D1, D2 are providing the current path for negative input voltage. 78
Figure 4.47 Simulated MOS diode current and voltage outputs over input signal for multiple temperatures. 78
Figure 4.48 Pulse detection principle based on peak current (voltage) and variable threshold. 79
Figure 4.49 Circuit schematic and device sizes of the current sensing path. 81
Figure 4.50 Circuit schematic, device sizes and bias current of the voltage sensing path. 83
Figure 4.51 Circuit schematic of Schmitt Trigger. 84
Figure 4.52 Pulse selector schematic. 84
Figure 4.53 Pulse timing diagram. 85
Figure 4.54 Stainless steel enclosure with mounted PCB and hybrid circuit including the HIGHTECS ASIC. . . 86
Figure 4.55 Photograph of the customized high temperature evaluation board used during HIGHTECS ASIC characterization measurements. 87
Figure 4.56 Layout of the frequency signal conditioning unit [1 *mm* × 1.5 *mm*] integrated onto the fabricated HIGHTECS ASIC. 87
Figure 4.57 Micrograph of the bonded HIGHTECS ASIC fabricated in the X-FAB XI10 SOI process. The frequency signal conditioning unit is positioned on the left side the ASIC (in blue). 88
Figure 4.58 Block diagram of the HIGHTECS ASIC hardware & software test platform. 89

Figure 4.59 Measured output frequency (red dots) via ARINC and FPGA at 25°C shows a linearity value of R^2 = 0.9999999684 with a reference clock frequency of F_{ref} = 12.288 MHz. 91
Figure 4.60 Measured output frequency (red dots) via ARINC and FPGA at 235°C shows a linearity value of R^2 = 0.9999999684 with a reference clock frequency of F_{ref} = 12.288 MHz. 92
Figure 4.61 Measured linearity values (R^2) of the output frequency over the 25°C to 235°C temperature range are within specification limits. The reference clock frequency value is F_{ref} = 12.288 MHz. 93
Figure 5.1 HIGHTECS ASIC in PGA package connected to ARINC 429 data reader. 96
Figure 5.2 ARINC 429 output from HIGHTECS ASIC. 97
Figure 5.3 ADC linearity plot of 2nd version of HIGHTECS ASIC assembled in PGA packages. 99
Figure 5.4 Characterisation board for testing of HIGHTECS ASIC in PGA package. 100
Figure 5.5 Voltage bandgap change with temperature and effective temperature coefficient of 2nd version of HIGHTECS ASIC. 101
Figure 5.6 Output from SG2 sensor on HIGHTECS module at +225°C. 102
Figure 5.7 Prototype SiC TVS devices with copper tags attached. 102
Figure 5.8 SEM picture of unbonded bond pad of adhesive bonded SOI device after 11,088 hours exposure to 250°C showing growth of whiskers. 104
Figure 5.9 Example of full day equivalent running for Profile 1. 105
Figure 5.10 Example of full day equivalent running for Profile 2. 105
Figure 5.11 Equipment for rapid change of temperature from –40°C to +225°C with 320 second cycle time. . . 106
Figure 5.12 Measured temperature profile for rapid change of temperature with 320 second cycle time. 106
Figure 5.13 HIGHTECS ASIC in PGA package connected to ARINC 429 data reader. 109

Figure 5.14 ARINC 429 output from HIGHTECS ASIC. 110
Figure 5.15 Cracking of die attach material after 375 cycles from –40°C to +250°C. 112
Figure 5.16 Test Box for HIGHTECS hybrid circuit and module. 113
Figure 5.17 Test of HIGHTECS hybrid circuit. 114
Figure 5.18 Test of HIGHTECS module. 114
Figure 5.19 Output from ARINC 429 monitor from HIGHTECS hybrid. 115
Figure 5.20 Tfo1 sensor output from HIGHTECS hybrid. . . . 116
Figure 5.21 Pulse generators used for testing of Qfreq sensor. 117
Figure 5.22 Qfreq sensor output against input frequency at room temperature. 118

List of Tables

Table 1.1 Clean Sky1 and Clean Sky2 targets 7
Table 3.1 Functional blocks for HIGHTECS ASIC 22
Table 4.1 Rail-to-Rail opamp corner simulation results . . . 31
Table 4.2 Corner simulation results of the PMOS input opamp with class AB output stage 36
Table 4.3 Process variation 38
Table 4.4 PVT corner simulation results of the NMOS input class AB output stage opamp. Process variation corners are presented in Table 4.3 38
Table 4.5 Bandgap voltage generator simulation results . . . 40
Table 4.6 Simulation results 43
Table 4.7 Simulation results of the voltage reference generator 46
Table 4.8 Specification versus achieved performance 57
Table 4.9 Functional blocks included in the ASIC 71
Table 4.10 Specification versus measurement results 92
Table 5.1 Summary of environmental tests on HIGHTECS ASIC in PGA package 100
Table 5.2 Lightning induced transient susceptibility – pin injection tests 103
Table 5.3 Summary of environmental tests carried out on SOI test chip 103
Table 5.4 Temperature storage tests at 200°C on HIGHTECS ASIC in PGA package 111
Table 5.5 Temperature storage tests at 250°C on HIGHTECS ASIC in PGA package 111
Table 5.6 Temperature cycling tests from –40°C to 250°C on HIGHTECS ASIC in PGA package 112
Table 5.7 Vibration and shock tests on HIGHTECS ASIC in PGA package 113

Table 6.1 Estimate of operating lifetime after extrapolation of temperature storage results for 1000 hours at 200°C and 250°C 120
Table 6.2 Summary of values derived from FMEA on HIGHTECS module 120
Table 6.3 Breakdown of weight by component for prototype HIGHTECS module 120
Table 6.4 Target and actual dimensions for prototype HIGHTECS module 121
Table 6.5 Target and actual current power consumption for prototype HIGHTECS module 121

List of Abbreviations

A/D	Analogue to Digital
ARINC	Aeronautical Radio Incorporated
ASIC	Application Specific Integrated Circuit
BJT	Bipolar Junction Transistor
CMOS	Complementary Metal Oxide Semiconductor
DC	Direct Current
DO-160	Environmental Conditions and Test Procedures for Airborne Equipment
ECU	Engine Control Unit
EHMS	Engine Health Management System
EMI	Electro-magnetic Interference
ESD	Electrostatic Discharge
FADEC	Full Authority Digital Engine (or Electronic) Control
GDP	Gross Domestic Product
HTCC	High Temperature Co-Fired Ceramic
JFET	Junction Gate Field Effect Transistor
MEMS	Microelectromechanical system
MOSFET	Metal Oxide Semiconductor Field Effect Transistor
PCB	Printed Circuit Board
PGA	Pin Grid Array
SiC	Silicon Carbide
SOI	Silicon on Insulator
VHDL	VHSIC Hardware Description Language

PART I

High Temperature Electronics Background

Lucian Stoica, Steve Riches, Colin Johnston

1

High Temperature Electronics for Aviation Applications

Aviation is a dynamic industry that continuously adapts to various market forces. The aviation market doubles in size, every 10 to 15 years, so there will be a greater need in the future for large aircrafts.

Key market forces that impact the airline industry are fuel prices, economic growth and development, environmental regulations, infrastructure, market liberalization, airplane capabilitics, other modes of transport, business models, and emerging markets [1]. Each of these forces can have both positive and negative impacts on the industry.

While the world economy GDP is expected to grow by 3.2% between 2012 to 2032, the number of airline passengers and airline traffic is expected to grow by 4.1% and 5%, respectively in the same interval.

The fleet size is expected to roughly double from 2013 to 2032 [1]. A long-term demand of 35280 new airplanes, valued at $4.8 trillion is forecasted [1]. 14350 of them will replace older, less efficient airplanes, reducing the cost of air travel and decreasing carbon emissions.

Europe is forecasted to be second largest market in the world by 2032 [1]. As shown in one of following section, from 2008 EU has already started to address and shape future aviation needs in Clean Sky and Clean Sky2 programs.

1.1 Value Story

Air traffic contributes today about 3% to global greenhouse gas emissions, and it is expected to triple by 2050 [2]. Although, other sectors are more polluting (electricity and heating produces 32% of greenhouse gases), pollution from air traffic is released high in the atmosphere where the impact is much greater. Meeting the climate and energy objectives will require reducing drastically the sector's environmental impact by reducing its emissions. Maximizing fuel efficiency to use less to go farther is also a key cost-cutting factor in a very competitive industry – and as air traffic increases, better noise reduction technologies are needed. Game-changing innovation in Aviation is risky, complex and expensive, and requires long-term commitment. This is why all relevant aviation stakeholders must work together to develop proof-of-concept demonstrators.

1.2 Fuel Prices Are Challenging the Airliners Profitability

Volatile oil prices have been the greatest challenge to airline profitability apart from the weak economy. *Fuel costs have surpassed labor as the largest segment of airline operating cost [1].* Fuel costs, approximately 13 percent of total costs in 2002, are closer to 34 percent today. After spiking in early 2012, oil prices have decreased in 2015. On the demand side, the weak economic outlook has moderated near-term growth projections. On the supply side, rising shale oil production in the United States is moderating near-term price projections. Lower jet fuel prices, are bolstering near-term airline profitability as shown in Figure 4 of [3]. However, long term projections for jet fuel are indicating a significant price increase [4], from approximately $60/barrel in 2015 to $90/barrel in 2020, $142/barrel in 2030 and $229/barrel in 2040. Jet fuel price is growing faster than other goods and services.

Therefore, there is a strong need for long term investment in the development of low consumption technologies for jet engines.

1.3 Growing Fuel Efficiency

Fuel costs have nearly doubled over the past 10 years. ***Fuel represents up to 30 percent of total operating cost for single-aisle airplanes and up to 50 percent for widebody airplanes [1]***. Fuel saving is a constant research topic of airplane manufacturers [5, 6], as this has a direct impact on costs. The main ways to save fuel for aero engines are presented in Figure 1.1 [5–7]. They must be balanced against all the costs and can only be realized when the initiative is fully deployed and sustained.

Airlines can improve their fuel efficiency in different ways [5–7]:

1. Deploying more fuel-efficient engines: replace older, less efficient airplanes with new-technology airplanes, such as the Boeing 787 or Airbus350 XWB. Weight reduction can be achieved by using composites and advanced avionics. Airbus has reported an 11% fuel burn improvement of A330neo versus current A330 at powerplant level [6].

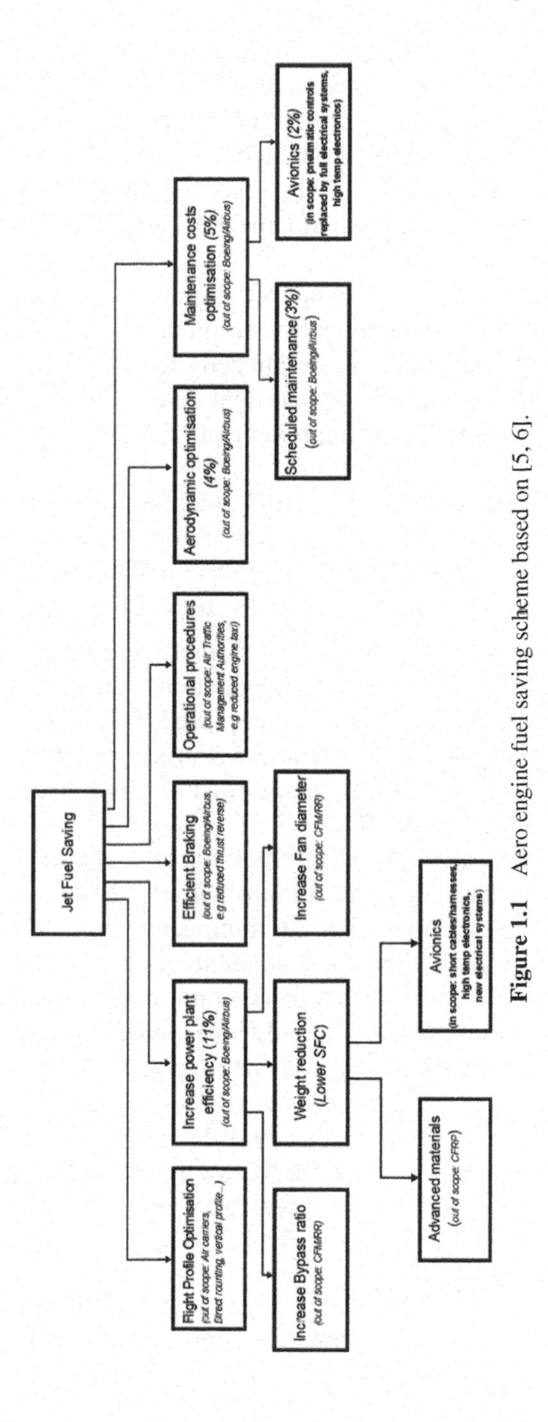

Figure 1.1 Aero engine fuel saving scheme based on [5, 6].

2. Improving operational procedures. Airlines can optimize fuel efficiency by making changes in operations, such as reducing the engine taxi time and the use of Auxiliary Power Unit (APU). Air carriers are also keen on raising the load factors on flights, which means making sure flights are close to or at aircraft capacity (all the seats are filled) [5].
3. Increasing braking efficiency by reducing the flap approach and a reduced thrust reverse [5].
4. Optimization of flight profile includes the optimum cruise altitude, the optimum climb/descent and the optimization of the cruising speed [5].
5. Optimization of aerodynamics & weight body shape by using of sharklets at the tips of the wings and the use of light composite materials. Airbus has reported an 4% fuel burn improvement of A330neo versus current A330 [6].
6. Maintenance costs optimization: Airbus has reported a 5% fuel burn improvement of A330neo versus current A330 due to lower direct maintenance costs [6]. This was achieved with longer maintenance intervals and by replacing the pneumatic controls with an electrical bleed air system.

The scope of current project is to focus on improving jet fuel saving by increasing the engine efficiency through a reduction of its weight which can be achieved with high temperature electronics placed closer to the engine such shortening the length of cables and harnesses. Further possible applications of high temp electronics, includes the replacement of pneumatic/mechanical controls with full electrical systems.

Benefits for high temperature electronics for aero-engines: By placing the electronics near to the sensor, the weight will also be reduced since the physical length between terminals will be minimized while the cost will be reduced since fewer cables will be needed, and the associated time to mount them on the engine will be also reduced. The fault rate will decrease as the signal is digitized before transmission and cables length is reduced. Sensor accuracy is improved as signal is digitized on the spot, also as cables length is minimized there is less noise coupling area to the signals. As the components are operational at higher temperatures there will be a reduced need for cooling. The flexibility of the system is increased as the components may be now placed in hot areas, which were previously inaccessible.

1.4 Clean Sky Initiative

The EU has taken a lead in green aviation technologies through Clean Sky1 and Clean Sky2 [8].

The Clean Sky Joint Technology Initiave started in 2008, and constitutes an industry wide, coherent program totaling €1.6 bn, equally shared by the EU and the European Aeronautical Industry.

Clean Sky2 is a natural continuation to progress achieved in Clean Sky1 (which has ended).

Clean Sky1 and Clean Sky2 are targeting very significant environmental gains, as shown in Table 1.1 [8].

By 2050, 75% of the world's fleet now in service (or on order) will be replaced by aircraft that can deploy Clean Sky2 technologies. Based on the same methodology applied in the Clean Sky1 economic case in 2007, the market opportunity related to these programes is estimated at ~€2000 bn. The direct economic benefit is estimated at ~€350–€400 bn and the associated spill-over is of the order of €400 bn.

The environmental case for continuing Clean Sky1 is even more compelling with an estimate of the CO_2 saving potential of 4 billion tones through Clean Sky2. These 4 billion tones of CO_2 to be saved from 2020 to 2050 will be additive to the approximately 3 billion tones achievable as a consequence of the Clean Sky Program.

GE was represented in Clean Sky1 by GE Aviation Systems (UK) & GE Global Research Munich as participants in High Temperature Survival Electronic Devices for Engine Control Systems (HIGHTECS) project no. 255749 working with Oxford University Materials.

1.4.1 Benefits of High Temperature Electronics for Jet Engine Controls and Health Monitoring

Environmental benefits: lower emissions – CO_2 reduction by 15–20%
For the aero-engine market, the extended high temperature electronics capability will facilitate the implementation of distributed architectures, where smart actuators and sensors can replace (or off-load) the centralised control

Table 1.1 Clean Sky1 and Clean Sky2 targets

	Clean Sky1*	Clean Sky2*
CO_2 and Fuel Burn	–20% to –40% (2020)	–20% to –30% (2025/2035)
NO_x	60% (2050)	–20% to –40% (2025/2035)
Population exposed to noise/Noise footprint impact	10dB to 20dB less noise (2020)	Up to –75% (2035)

* = Baseline for Clean Sky1 and Clean Sky2 figures are best available performance in 2000 and 2014, respectively.

electronics. Up to 500 conductors are currently used for interfacing between jet engine sensors, actuators, flight control computers and the centralised FADEC. The application of distributed architectures could reduce the conductor count from 500 to 8 for duplex control, offering cable and harness weight saving, connector pin reduction, fault reduction and a simpler FADEC [9]. This type of electronic unit would be installed inside the actuator or sensor housing and would consist of the sensor signal conditioning electronics, A/D converters, multiplexers and a serial interface bus [9].

At present, long, high-temperature mineral insulated (MI) or fibre-optic cable is required to connect the sensor to the electronics located in a more benign region of the gas turbine. Electronics co-located with the sensor will lead to a reduction in associated cabling, connectors, and terminals leading to reductions in weight and parts count (hence cost). The development of MEMS sensors with electronics integrated onto a multi-chip module could also lead to significant enhancement of performance at reduced costs. Moreover, the ability to perform signal handling/conditioning prior to engine control unit (ECU) will have benefits in terms of enhancing the data available for engine health monitoring. For example, temperature signals from thermocouple arrays must be averaged prior to sending the signal to the ECU as weight restrictions do not allow for individual cables from each thermocouple to be relayed to the ECU. The use of a multiplexing systems that can withstand engine casing temperatures (~250°C) would allow individual thermocouple signals to be analyzed by the ECU off a single cable harness. This could permit the detection of engine hotspots, radial distortions in temperature and condition monitoring of individual thermocouples.

Managing engine performance is receiving a greater amount of attention for safety, reliability and fuel burning savings [10]. Advances in heat resisting sensors and the desire to use full authority digital control electronics (FADEC) and engine health monitoring systems (EHMS) near to the sensing element is accelerating the interest in the use of high temperature electronics. This is leading to the development of "intelligent sensors", which incorporate high temperature electronics in the sensor itself and have the capability to perform self-diagnosis of their health. The output of the "intelligent sensor" will be a digital signal which is then fed into the FADEC. The reduced need for processing of analogue signals within the FADEC unit can increase the capacity for incorporating the EHMS within the same unit, saving weight, space and costs.

For the aerospace market, improved sensor technology will have significant benefits in a number of areas. Firstly, although sensor weight may be small relative to the total weight of the aircraft, any improvements that could be achieved through reductions in lead-outs, terminals, connectors, etc. can still have a tangible impact on fuel consumption and running costs. For example, weight savings of even a few kilos can result in hundreds of thousands of pounds in annual fuel saving. Secondly, improved engine monitoring capability should result in engines being run at conditions for more optimal thermodynamic efficiency, resulting in reduced fuel consumptions (and engine emissions) and potentially increased component life. Moreover, improved sensor performance could lead to a reduction in maintenance costs through "smart scheduling" of servicing and overhaul based on reliable and indicative sensor data and not on fixed flight hour intervals.

References

[1] *Current Market Outlook 2013–2032*, Boeing, www.boeing.com.
[2] *Overview of Clean Sky 2 Initiative*, October 2013.
[3] *Economic Performance of the Airline Industry*, IATA, Brian Pearce, Chief Economist.
[4] *Annual Energy Outlook 2015, with projections to 2040*, DOE/EIA – 0383 (2015), April 2015.
[5] *Airbus Customer Service, Fuel and Emissions Performance Manager, Simon Weselby, Saving Fuel: It's A Team Sport*, IATA Maintenance Cost Conference, October 2012.
[6] *Airbus, The A330neo Powering into the future*, John Leahy, Chief Commercial Officer – Customers.
[7] *US Energy Information Administration*, www.eia.gov/todayinenergy/detail.cfm?id=6670
[8] *Overview of the Proposed Programme, Clean Sky 2*, October 2013.
[9] *HIGHTECS Project Final Report*, Steve Riches, 26 October 2012.
[10] *HIGHTECS Cahier Des Charges Techniques – Technical Specification v3.0*, Turbomeca, Groupe Safran, 2011.

2

High Temperature Integrated Technologies

2.1 Introduction

There is a growing desire to install electronic power and control systems in high temperature environments to improve the accuracy of critical measurements, reduce the amount of cabling and elimination of cooling systems. Typical applications include down-hole petroleum/gas/geothermal exploration and production, turbine engines for aircraft propulsion and power generation and power modules for electric/hybrid vehicles [1–4].

Fuel costs for aeroengines have approximately doubled over the past 10 years and now represent up to 50% of the operating costs of many modern widebody aircraft [5]. Reducing specific fuel consumption by reducing aircraft weight has become a major focus for research and development. The use of sensors developed to operate for long periods in high temperature environments allows sensors to be replaced close to the engine sensing and control units eliminating the need for complex heat sinks, special fuel pumping and interfacing, which in turn assists with the goal of aircraft weight reduction [6]. Mounting the engine sensing and control unit close to the sensors means ambient temperatures may easily reach 200°C. This requirement has posed a challenge to the bulk CMOS technologies which are typically qualified for operation between –55°C and 125°C. The leap in operating temperature to above 200°C in combination with high pressures, vibrations and potentially corrosive environments means that different semiconductors, passives, circuit boards and assembly processes will be needed to fulfill the target performance specifications. Although extensive research to investigate temperature related reliability effects in semiconductors such as leakage current, electromigration and time dependent dielectric breakdown (TDDB) has been carried out [7], understanding the design constraints, development of robust packaging systems and reliable interconnections are the key to the success of high temperature electronics systems. The main advantage of SOI technology in high temperature applications are the reduced leakage current due to the reduced junction area and reduced latchup due to isolated PMOS (P-type Metal Oxide Semiconductor Logic) and NMOS (N-type Metal Oxide Semiconductor Logic) transistors [8]. CMOS SOI technology has been shown to be better suited for high temperature operation over bulk CMOS [9–11]. The reliability of CMOS SOI for use at 250°C was presented in [12]. CMOS SOI integrated circuits have been designed and tested for high-temperature applications up to 300°C in [13–19] and up to 400°C in [20]. Silicon carbide (SiC) BJT, JFET and MOSFET based integrated circuits have been demonstrated up to 600°C [21–23]. High temperature electronics technologies and applications have been recently reviewed in [24]. The major effects of elevated temperature on semiconductor material and devices are:

- An exponential increase in leakage current of reverse-biased *pn* junctions. This might significantly limit the performance of bulk-CMOS ICs, where the transistors are isolated from the common bulk by means of a *pn* junction. In SOI technologies, however, buried oxide prevents any leakage current into the bulk, thus making this technology well suited for high temperature applications. Still, structures necessarily incorporating *pn* diodes, like ESD protection circuits, may adversely influence the performance of a system at high temperatures.
- Carrier mobility degradation, occurring with the rate of T^{-n} for MOS devices, where *n* ranges from 1.5 to 1.8 between 25°C and 200°C [17]. This directly impacts the transfer characteristics of MOS transistors since drain current of the saturated devices is proportional to the carrier mobility in the channel.
- Finally, the threshold voltage shifts by 1–3 $mV/^{\circ}C$ as the Fermi potential, the depletion width and charge under channel reduce with temperature [17].

The above mentioned temperature effects were accounted for during the design of the circuits presented in this book by using the Zero-Temperature Coefficient (ZTC) and "g_m/I_d" methodologies [25].

The European Union (EU) has taken a lead in green aviation technologies by funding projects such as Clean Sky1 and Clean Sky2 [26]. The Clean Sky Joint Technology Initiave started in 2008, and constitutes an industry wide program targeting very significant environmental gains: a reduction of CO_2 and NO_x emissions of 40% and 60%, respectively. General Electric was represented in Clean Sky by GE Aviation Systems (UK) and GE Global Research Munich as participants in the High Temperature Survival Electronic

Devices for Engine Control Systems (HIGHTECS) project working with Oxford University Materials.

The HIGHTECS design concept was to take the output from several on-engine sensors (temperature probe, thermocouple, strain gauges, frequency) and carry out the signal conditioning on the sensor signals, multiplexing, analogue to digital conversion, and transmission of the data through a serial data bus on a single ASIC (Application Specific Integrated Circuit) [6]. The unit was designed to meet the environmental requirements of DO-160 for a helicopter engine, with the specific needs of operation at 200°C with a lifetime of 50,000 engine operating hours. Due to the temperature and lifetime requirements, and the current feasibility of SOI technology over SiC, the HIGHTECS ASIC was fabricated as a custom CMOS SOI device to be assembled on a ceramic hybrid carrier [6, 27]. The hybrid was assembled in a stainless steel enclosure, mounted on an aeroengine during tests on the ground, and due to the shorter cables needed in between the sensors and the electronics, it helps reducing the weight of the aeroengine by several kilograms [28].

References

[1] B. Parmentier, O. Vermesan, and L. Beneteau, “Design of high temperature electronics for well logging applications,” in *Proc. International Conference on High Temperature Electronics (HiTEN)*, Oxford, United Kingdom, Jul. 2003, pp. 77–84.

[2] B. Ohme and M. Larson, “Analog component development for 300°C sensor interface applications,” in *Proc. International Conference on High Temperature Electronics (HiTEC)*, Albuquerque, United States, May. 2012, pp. 1–17.

[3] D. MacGugan, “DM300 – a 300°C geothermal directional module development,” in *Proc. International Conference on High Temperature Electronics (HiTEC)*, Albuquerque, United States, May. 2012, pp. 293–300.

[4] R. W. Johnson, J. L. Evans, P. Jacobsen, J. R. Thompson, and M. Christopher, “The changing automotive environment: High-temperature electronics,” *IEEE Trans. Electron. Packag. Manuf.*, vol. 27, pp. 164–176, Jul. 2004.

[5] Boeing. (2014, Sep.) Current market outlook 2013–2032. [Online]. Available: www.boeing.com/boeing/commercial/cmo/

[6] L. Stoica, V. Solomko, T. Baumheinrich, R. D. Regno, R. Beigh, S. Riches, I. White, G. Rickard, and P. Williams, "Design of a high temperature signal conditioning ASIC for engine control systems – HIGHTECS," in *Proc. IEEE International Symposium on Circuits and Systems*, Melbourne, Australia, May. 2014, pp. 2117–2120.

[7] J. D. Cressler and H. A. Mantooth, *Extreme Environment Electronics*. CRC Press, 2012.

[8] D. Vanhoenacker-Janvier, M. E. Kaamouchi, and M. S. Moussa, "Silicon-on-insulator for high-temperature applications," *IET Circuits Devices Syst.*, vol. 2, pp. 151–157, Feb. 2008.

[9] G. Shahidi, "Mainstreaming of the SOI technology," *Proceedings SPIE Microelectronic Device Technology III*, vol. 3881, Oct. 1999.

[10] P. Francis, A. Terao, B. Gentinne, D. Flandre, and J.-P. Colinge, "SOI technology for high-temperature applications," in *Proc. IEDM Tech. Dig.*, San Francisco, United States, Dec. 1992, pp. 13.5.1–13.5.4.

[11] D. Flandre, A. Nazarov, and P. Hemment, Eds., *Science and Technology of Semiconductor-On-Insulator Structures and Devices Operating in a Harsh Environment*. Dordrecht, The Netherlands: Kluwer Academic Publishers, 2005.

[12] K. Grella, S. Dreiner, H. Vogt, and U. Paschen, "Reliability of CMOS Silicon-in-Insulator for use at 250°C," *IEEE Trans. Device Mater. Rel.*, vol. 14, pp. 21–29, Mar. 2014.

[13] J. Eggermont, D. D. Ceuster, D. Flandre, B. Gentinne, P. Jespers, and J. Colinge, "Design of SOI CMOS operational amplifiers for applications up to 300°C," *IEEE J. Solid-State Circuits*, vol. 31, pp. 179–186, Feb. 1996.

[14] D. Flandre, S. Adriaensen, A. Afzalian, J. Laconte, D. Levacq, C. Renaux, L. Vancaillie, J. Raskin, L. Demeus, P. Delatte, V. Dessard, and G. Picun, "Intelligent SOI CMOS integrated circuits and sensors for heterogeneous environments and applications," in *Proc. IEEE Sensors*, Orlando, United States, Jun. 2002, pp. 1407–1412.

[15] J. O'Connor, J. Tsang, and J. McKitterick, "225°C high temperature silicon-on-insulator (SOI) ASICs for harsh environments," in *Proc. IEEE International Workshop for Integrated Power Packaging (IWIPP)*, Chicago, United States, Sep. 1998, pp. 2–5.

[16] H. Kappert, N. Kordas, S. Dreiner, U. Paschen, and R. Kokozinski, "High temperature SOI CMOS technology and circuit realization for

applications up to 300°C," in *Proc. IEEE International Symposium on Circuits and Systems*, Lisbon, Portugal, May. 2015, pp. 1162–1165.

[17] L. Demeus, V. Dessard, A. Viviani, S. Adriaensen, and D. Flandre, "Integrated sensor and electronic circuits in fully depleted SOI technology for high-temperature applications," *IEEE Trans. Ind. Electron.*, vol. 48, pp. 272–280, Apr. 2001.

[18] F. Silveira, D. Flandre, and P. G. A. Jespers, "A gm/Id based methodology for the design of CMOS analog circuits and its application to the synthesis of a silicon-on-insulator micropower OTA," *IEEE J. Solid-State Circuits*, vol. 31, pp. 1314–1319, Sep. 1996.

[19] J. Watson, "An ultra-low noise instrumentation amplifier designed for high temperature applications," in *Proc. International Conference on High Temperature Electronics (HiTEC)*, Albuquerque, United States, May. 2012, pp. 82–86.

[20] A. Schmidt, H. Kappert, and R. Kokozinski, "High temperature analog circuit design in PD-SOI CMOS technology using reverse body biasing," in *Proc. IEEE 39th European Solid-State Circuit Conference*, Bucharest, Romania, Sep. 2013, pp. 359–362.

[21] P. Neudeck, D. Spry, L. Chen, G. Beheim, R. Okojie, C. Chang, R. Meredith, T. Ferrier, L. Evans, M. Krasowski, and N. Prokop, "Stable electrical operation of 6HSiC JFETs and ICs for thousands of hours at 500°C," *IEEE Electron Device Lett.*, vol. 29, pp. 456–459, May. 2008.

[22] R. Ghandi, C.-P. Chen, L. Yin, X. Zhu, L. Yu, S. Arthur, F. Ahmad, and P. Sandvik, "Silicon carbide integrated circuits with stable operation over a wide temperature range," *IEEE Electron Device Lett.*, vol. 35, pp. 1206–1208, Dec. 2014.

[23] S. Garverick, C.-W. Soong, and M. Mehregany, "SiC JFET integrated circuits for sensing and control at temperatures up to 600°C," in *Proc. IEEE Energytech*, Cleveland, United States, May. 2012, pp. 1–6.

[24] J. Watson and G. Castro, "A review of high-temperature electronics technology and applications," *Journal of Material Science: Materials in Electronics*, vol. 26, pp. 9226–9235, Jul. 2015.

[25] D. Flandre, L. Demeus, V. Dessard, A. Viviani, B. Gentinne, and J. P. Eggermont, "Design and application of SOI CMOS OTAs for hightemperature environments," in *Proc. IEEE 24th European Solid-State Circuit Conference*, The Hague, The Netherlands, Sep. 1998, pp. 404–407.

[26] CleanSky. (2013, Oct.) Clean Sky2 overview of the proposed programme. [Online]. Available: http://www.cleansky.eu/content/document/clean-sky-2-overviewproposed-programme
[27] S. Riches, C. Warn, K. Cannon, G. Rickard, and L. Stoica, "Design and assembly of high temperature distributed aero-engine control system demonstrator," in *Proc. International Conference on High Temperature Electronics (HiTEC)*, Albuquerque, United States, May. 2014.
[28] J. Rigaud and L. Stoica, "Turbomeca TCON," Oct. 2015, Meeting Notes.

PART II

Development of Multi-Sensor Data Acquisition System

Lucian Stoica, Valentyn Solomko, Ozan Iskilibli,
Renato Del Regno, Reece Beigh, Thorsten Baumheinrich,
Steve Riches, Colin Johnston, Geoff Rickard, Paul Williams

3

Outline of System

3.1 High Level Input Specification

A high level draft technical specification was provided by Turbomeca as the basis for the design of the high temperature electronics platform. The design concept was to take the output from several on-engine sensors (temperature probe, thermocouple, strain gauges, frequency), carry out the signal conditioning on the sensor signals, multiplexing, analogue to digital conversion and transmission of the data through a serial data bus. The DC power supply for the unit is provided by the FADEC. The unit has to meet the environmental requirements of DO-160 for a helicopter engine, with the specific need to operate at 200°C, with short term operation at temperatures up to 250°C. The system service lifetime target is 50,000 engine flight hours.

3.2 Technology Assessment and Selection

A review of the options for the high temperature electronics to be considered for the HIGHTECS module was carried out. This review included the availability of devices and components, the status of high temperature electronics packaging technology, an assessment of the technology maturity, potential failure modes and a review of accelerated life tests to predict service life.

For the electronic devices and components, an ASIC based on a Silicon-on-Insulator (SOI) semiconductor manufactured using the X-FAB 1 μm SOI foundry in Germany was selected to perform the analogue signal conditioning, multiplexing, ADC (Analogue to Digital Conversion), logic control and serial data transmission. The circuit also required additional high temperature voltage regulators, a clock oscillator, capacitors, precision resistors and lightning protection devices, all of which should be capable of meeting the high temperature operating conditions. The review highlighted the limitations of ceramic based capacitors and Si based lightning protection devices. High temperature silicon capacitors produced by Ipdia – France became available during the course of the project and development SiC transient voltage suppressors, which have potential for operation above 150°C were evaluated.

The status of high temperature electronics packaging for the HIGHTECS module was also reviewed, covering materials and processes for die attach and wire bonding, attachment of passive devices and packaged components to ceramic substrates and connections for external inputs/outputs to/from the HIGHTECS module. An assessment of potential failure modes relating to the packaging technology options was undertaken, which highlighted areas to focus on within the testing programme.

The review covered the use of accelerated reliability tests to predict service life. In conclusion, the following tests were defined to address the concerns for the reliability of the electronics components and packaging technology operating at high temperature:

- Long term temperature storage at +250°C to assess the long term degradation at temperature
- Rapid thermal cycling from –40°C to +225°C to represent the stresses endured during the typical flight profile
- Vibration at room temperature and at 200°C to investigate whether the combined effect of vibration and temperature accelerates any failure mechanism

Tests have been carried out to investigate these factors on a SOI test chip.

3.3 Definition of Prototype System

The design principle of the HIGHTECS module was based on a custom silicon on insulator (SOI) ASIC being used for the majority of the signal processing and conditioning from the range of sensors (i.e. temperature probe, strain gauges, thermocouple, frequency), multiplexing, analogue to digital conversion and transmission of data through an ARINC 429 databus. The ASIC was then integrated with a high temperature external clock and packaged onto a ceramic hybrid circuit. This hybrid circuit was assembled in a Kovar package together with development high temperature SiC based transient voltage suppressors, which was hermetically sealed in an inert gas atmosphere. The Kovar package was then mounted into a stainless steel enclosure containing high temperature connectors and EMI shielding.

3.3.1 HIGHTECS SOI ASIC

The ASIC block diagram for the HIGHTECS module is presented in Figure 3.1 From the analogue sensor outputs (temperature probe, strain gauges, thermocouple), the signals pass through buffers/low pass filters for conditioning and then into a 10:1 analogue multiplexer. The output from the analogue multiplexer is fed to an analogue to digital converter which outputs to the ARINC 429 bus.

From the frequency outputs (Nfreq and Qfreq), the signals are processed using comparators/counters, synchronised with an external clock and sent to a 16b digital multiplexer. The DIN input is also sent to the digital multiplexer. The digital multiplexer outputs to dual ARINC 429 buses. The ARINC 429 bus is the selected serial output from the HIGHTECS module.

The functional blocks for the HIGHTECS ASIC are presented in Table 3.1.

A picture of the 1st version of the HIGHTECS ASIC is presented in Figure 3.2. The ASIC contains all the sensor conditioning circuits, ADC,

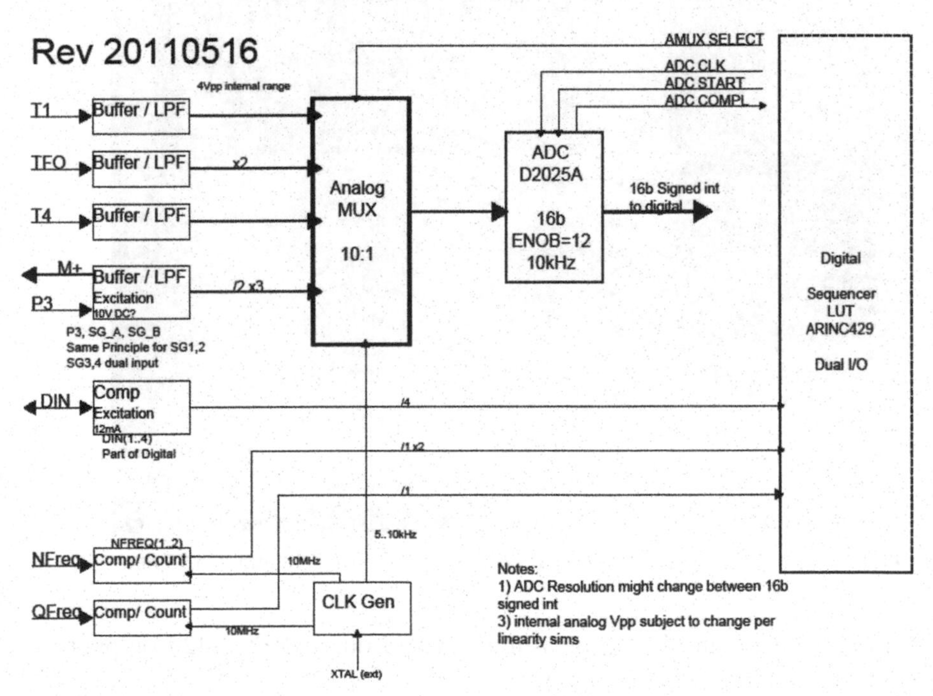

Figure 3.1 Block diagram for SOI ASIC in HIGHTECS module.

Table 3.1 Functional blocks for HIGHTECS ASIC

Block Name	Block Name
SG1	Bandgap
SG2	Global current mirrors
P3	Voltage generator
T4	Reference current generator
T1	ARINC Driver (x2)
TFo	DIN (4i/ps)
NFreq	ARINC Control sequencer
QFreq	Nfreq & Qfreq logic
ADC	

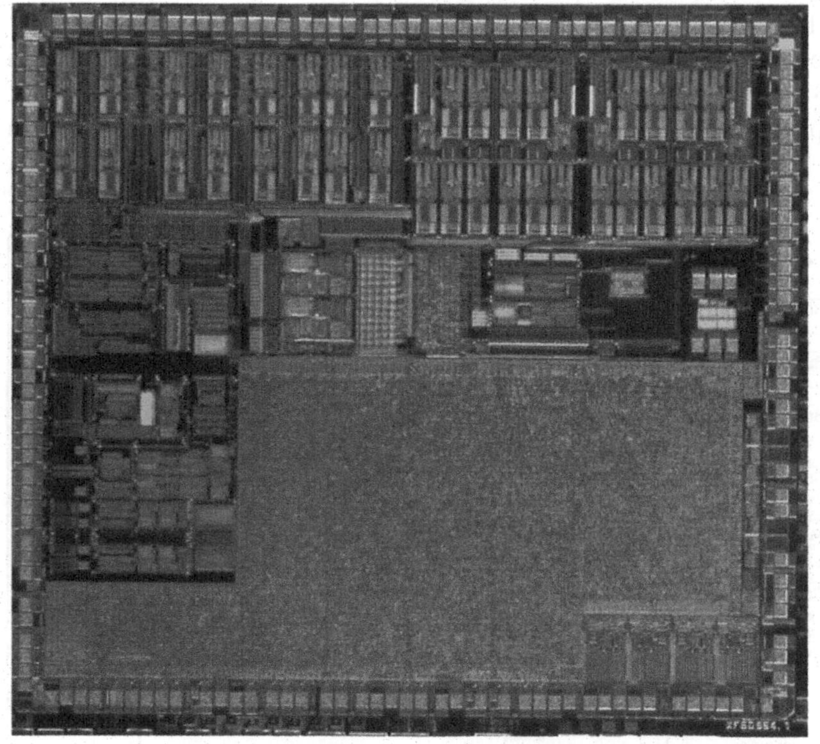

Figure 3.2 1st Version of HIGHTECS ASIC – device size 7.48 mm × 5.95 mm.

Multiplexer, Qfreq and Nfreq measurement and dual ARINC 429 outputs. The die size is 7.48 mm × 5.95 mm.

The 2nd version of the HIGHTECS ASIC was re-laid out and manufactured at XFAB, a die picture is shown in Figure 3.2. Modifications were made to the layout of the connections to the ADC including bringing out of voltage references, and changes to the VHDL code for Tfo2 and Nfreq.

3.3.2 HIGHTECS Hybrid Circuit

The HIGHTECS hybrid circuit layout is presented in Figure 3.3. The hybrid circuit contains the following components in addition to the HIGHTECS SOI ASIC:

- Voltage Regulators
- External Clock Generator/Crystal Oscillator
- Prototype SiC Transient Voltage Suppressors
- Resistors
- Capacitors

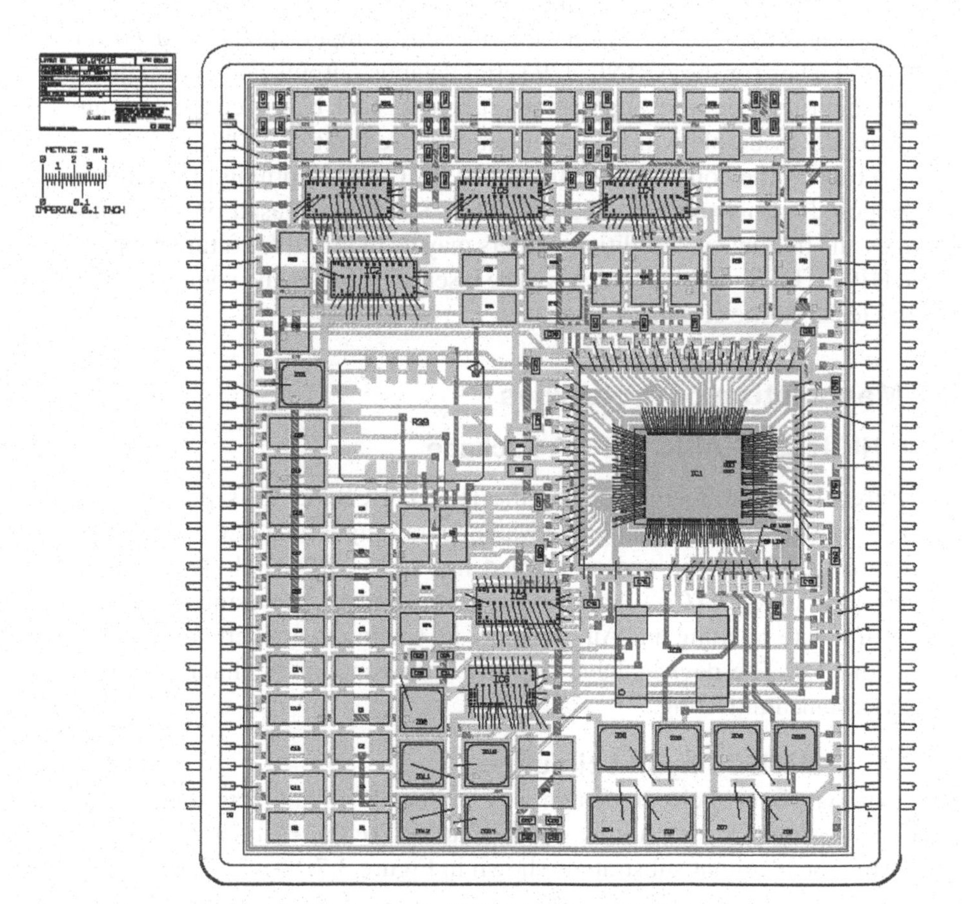

Figure 3.3 Layout of HIGHTECS hybrid circuit.

In addition to the hybrid circuit a high temperature printed circuit board containing resistors required for the frequency sensors was designed and manufactured. This board was also mounted in the HIGHTECS module.

3.3.3 HIGHTECS Module

A drawing of the HIGHTECS module assembly is presented in Figure 3.4.

The hybrid circuit (containing the ASIC) sealed in a hermetic Kovar package and the high temperature printed circuit board (containing resistors) is mounted into the stainless steel enclosure. A clamping plate is used to fix the Kovar package in place.

The stainless steel enclosure is completed by mechanical fixing of a lid with an EMI shielding gasket to the stainless steel base. Future versions may be welded, but, at this stage, a removable lid is preferred.

Two connectors are used; one for the sensor input signals and power supply, the other for the ARINC 429 serial databus outputs and connections. For the high temperature application, stainless steel based connectors are commercially available with an upper temperature limit of 260°C.

Filtering on the connector for improved EMC and lightning protection is not proposed at this stage for the HIGHTECS module. Based on the signal voltages and frequencies, additional filters may be required for future versions of the HIGHTECS module, which will be inserted between the connector pins and the leads on the Kovar package.

3.4 Manufacture of Prototypes

3.4.1 HIGHTECS ASIC in PGA Package

A picture of a Si wafer containing the HIGHTECS ASIC is presented in Figure 3.5. After initial probing, the wafer containing HIGHTECS ASIC was sawn into individual die and assembled into a 181 I/O High Temperature Co-Fired Ceramic (HTCC) Pin Grid Array (PGA) package using die attach, aluminium wire bonding and Au-Sn solder lid sealing in an inert atmosphere, see Figure 3.6. The devices have been used for functional, characterisation and environmental testing.

3.4.1.1 HIGHTECS hybrid circuit

The prototype hybrid circuit design was laid out for manufacture on a 96% alumina substrate. The circuit was built up using Au thick film and dielectric layers and the resultant substrate is shown in Figure 3.7.

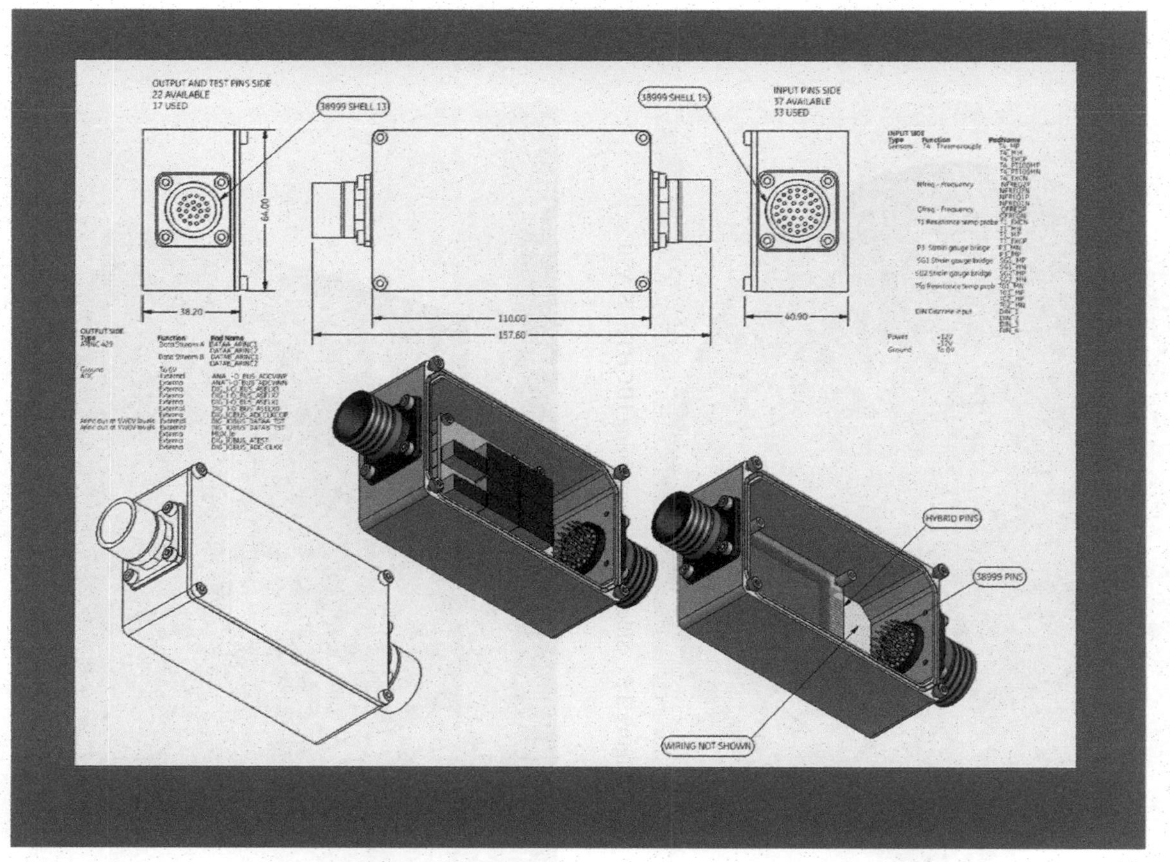

Figure 3.4 Mechanical assembly drawing for HIGHTECS module.

Figure 3.5 Silicon wafer containing HIGHTECS ASICs.

Figure 3.6 HIGHTECS ASIC assembled in HTCC PGA package.

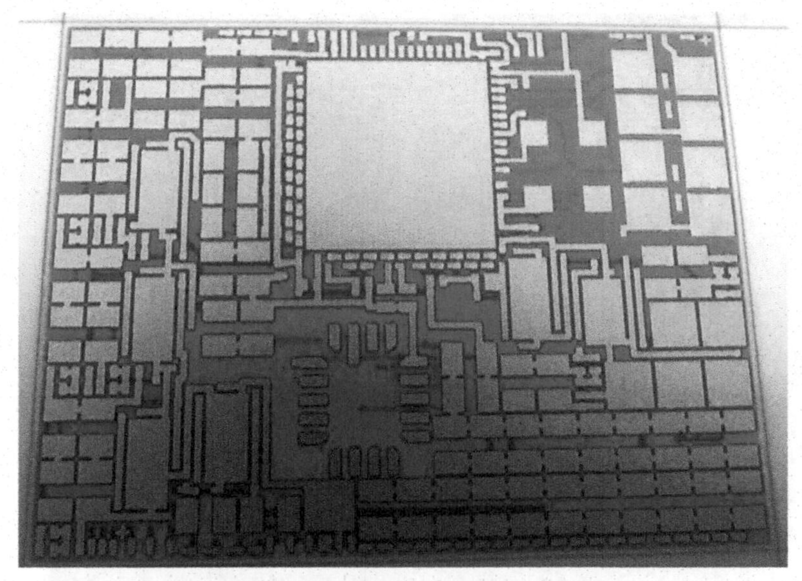

Figure 3.7 HIGHTECS hybrid circuit substrate.

The following components were assembled onto the substrate and the populated substrate is shown in Figure 3.8.

- HIGHTECS ASIC
- Interposer
- Voltage Regulators
- Clock Oscillator
- Precision Resistor
- Resistors
- Capacitors

The bare die components including the silicon capacitors were attached onto the thick film pads on the alumina substrate. TVS devices were only assembled into some of the hybrid circuits.

3.4.1.2 Assembly of Ceramic Substrate to Metal Package

The populated substrate was mounted into the metal package as shown in Figure 3.9. Some of these samples were populated with prototype SiC transient voltage suppressors.

A lid was resistance seam sealed onto the metal package in an inert atmosphere and gross/fine leak tested.

Figure 3.8 HIGHTECS populated hybrid circuit substrate.

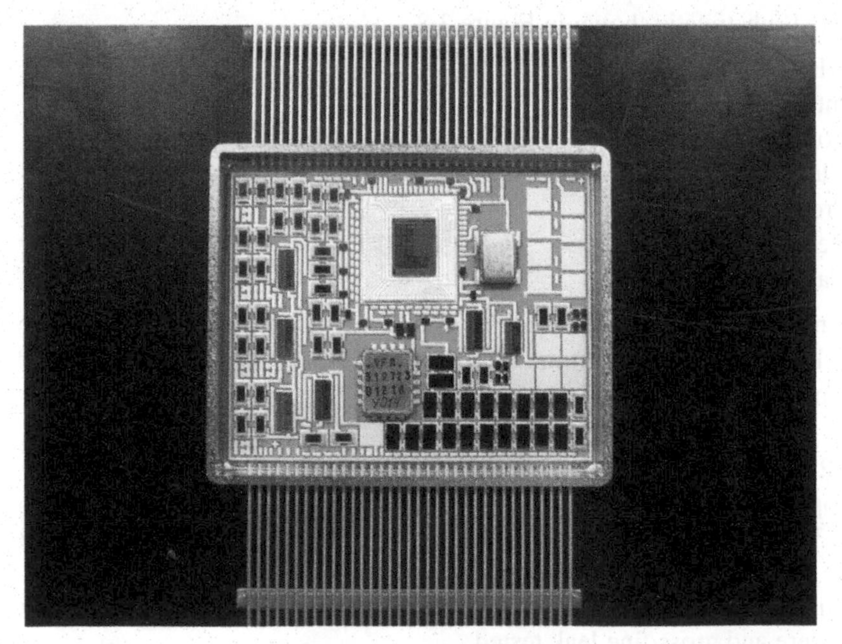

Figure 3.9 HIGHTECS hybrid circuit mounted in metal package.

3.4.1.3 High Temperature PCB for Resistors

The size of the derated high temperature resistors for the frequency circuits precluded their use in the hybrid circuit. A separate high temperature circuit board was designed specifically for these high temperature resistors and the components were assembled using a high melting point solder, as shown in Figure 3.10.

3.4.1.4 HIGHTECS Module

The hybrid circuit and high temperature PCB containing the resistors were mounted into the stainless steel enclosure, as shown in Figure 3.11.

The connections between the leads on the metal package, connection pads on the printed circuit board and the connectors in the stainless steel enclosure were made with polyimide insulated copper wire, attached with high melting point solder.

The HIGHTECS module with the removable lid attached is shown in Figure 3.12.

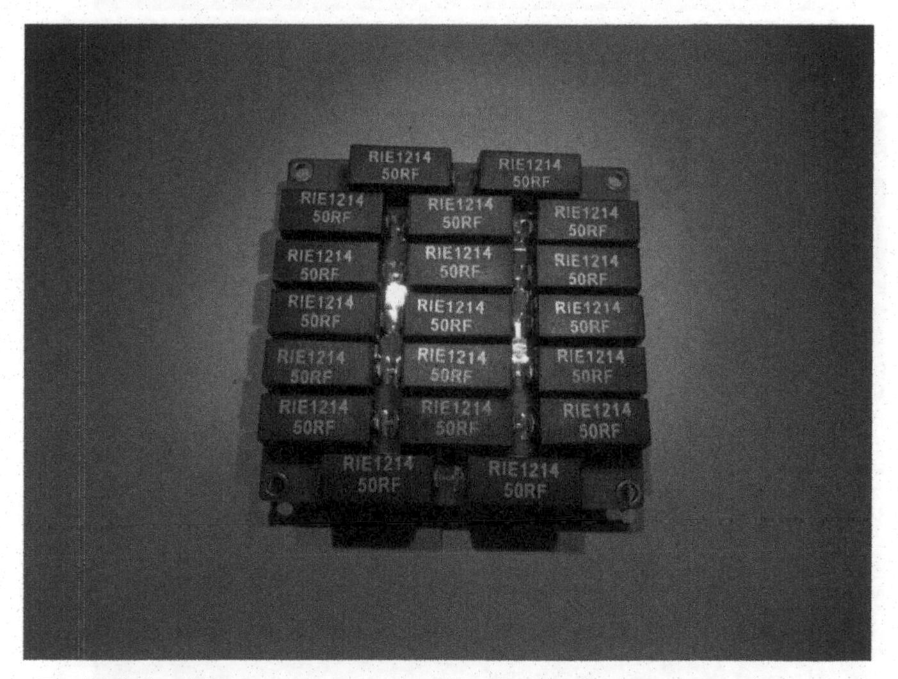

Figure 3.10 Resistors surface mounted onto high temperature printed circuit board.

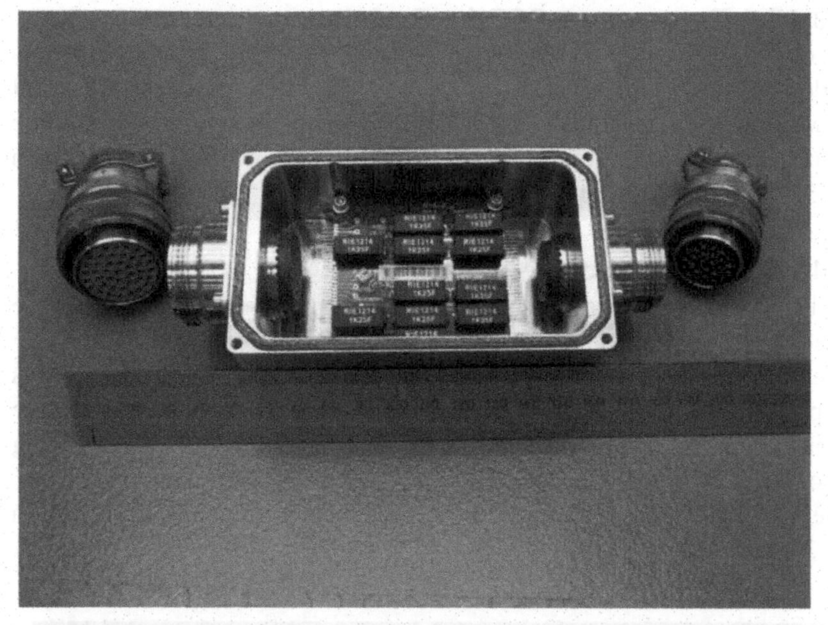

Figure 3.11 Stainless Steel enclosure with mounted PCB and hybrid circuit.

Figure 3.12 Stainless steel enclosure with lid incorporating EMI gasket.

4

Design and Characterization of HIGHTECS Signal Channels and Building Blocks

4.1 Operational Amplifiers

Three operational amplifiers with class-AB output stages were designed during HIGHTECS: one with a PMOS input stage, one with a NMOS input stage and one with a rail-to-rail input stage [1]. The rail-to-rail version is presented in Figure 4.1. The opamps have been used in the signal conditioning blocks as well as the bias block for stable generation of currents and voltages, and these, in turn, have been used to bias the strain, temperature, and frequency channels for signal conditioning.

The simulated performance of the rail-to-rail opamp across corners is presented in Table 4.1. While the opamps were not measured directly, their performances were indirectly evaluated in the signal conditioning instrumentation amplifiers and in the voltage to voltage and voltage to current converters in the bias block. These bias and signal conditioning channels were measured and the results will be shown in the measurement results section.

Table 4.1 Rail-to-Rail opamp corner simulation results

Specification	Min	Nom	Max	Units
Supply Voltage	4.5	5	5.5	V
Bias Current	30	55	70	μA
Current Consumption		586	860	μA
Temperature	–60	27	+225	°C
Input Voltage	0.5	2.5	4.5	V
Load Current	–100	50	100	μA
Capacitive Load	0	1.5	3	pF
DC open-loop Gain	64	106	113	dB
3-dB BW in Voltage-follower Config.	2.6	9.94	23.3	MHz
Phase Margin	52	72.78	99	o
Gain Margin	5.8	14.2	40	dB

4.1.1 Rail-to-Rail OpAmp

4.1.1.1 Schematic diagram

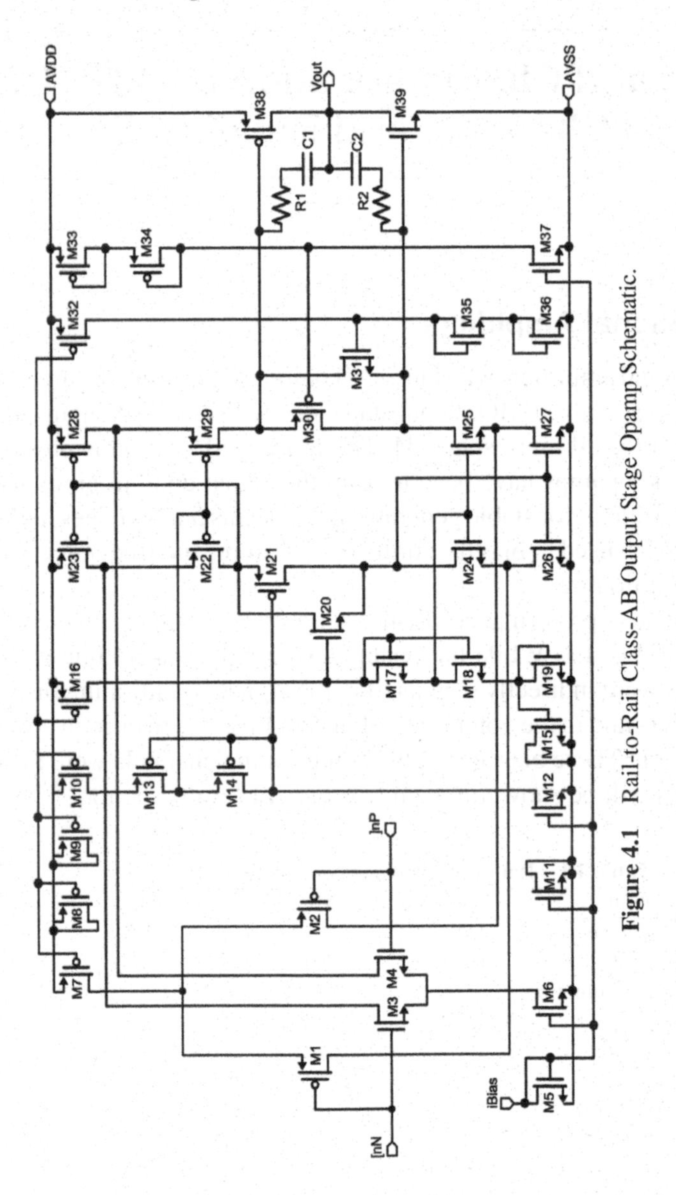

Figure 4.1 Rail-to-Rail Class-AB Output Stage Opamp Schematic.

4.1.1.2 Layout

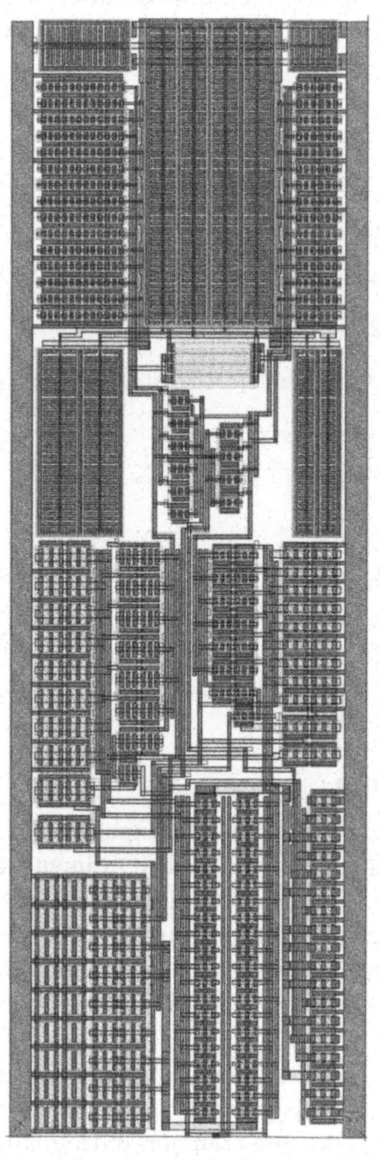

Figure 4.2 Rail-to-Rail Class-AB Output Stage Opamp Layout.

4.1.1.3 Simulation results

The OpAmp model for simulation includes layout capacitive and resistive parasitics. The design was simulated over 505 corners defined as combination of following parameters:

```
DesignVar = {'CL' '0p' '3p' 'x' 'x' 'x'; ...
            'Vdc' '0.5' '2.5' '4.5' 'x' 'x'; ...
            'Iref' '30u' '70u' 'x' 'x' 'x'; ...
            'IL' '-100u' '0u' '100u' 'x' 'x'; ...
            'temp' '-60' '225' 'x' 'x' 'x'};
ModelSection = {'bsim3v3.scs' 'cap.scs' 'res.scs' 'dio.scs'; ...
            'tm' 'tm' 'tm' 'tm' ; ...
            'wp' 'wp' 'wp' 'tm' ; ...
            'wp' 'ws' 'ws' 'tm' ; ...
            'ws' 'wp' 'wp' 'tm' ; ...
            'ws' 'ws' 'ws' 'tm' ; ...
            'wo' 'wp' 'wp' 'tm' ; ...
            'wz' 'wp' 'wp' 'tm' };
```

4.1.2 PMOS-input OpAmp

Several operational amplifiers with class-AB output stages [2] were designed: one with a PMOS input stage, one with a NMOS input stage and one with a rail-to-rail input stage [1, 3]. The PMOS input stage class-AB output stage version presented in Figure 4.3 have been used inside the instrumentation amplifier (IA) for the strain gauge signal conditioning channel and in the Single Ended to Differential Converter (SEDC) placed before the analogue to digital converter (ADC).

The simulated performance of the PMOS opamp across process, voltage and temperature (PVT) variation is presented in Table 4.2. The opamp has a PMOS input stage M_1, M_2, allowing common-mode input voltage down to the negative rail. Class-AB control is provided by M_{19} and M_{23} devices. The cascode mirror $M_{15}, M_{16}, M_{20}, M_{21}$ is biased by $M_{10}-M_{12}$ while the cascode mirror $M_{13}, M_{14}, M_{17}, M_{18}$ is being biased by M_5-M_7. A constant operating point over the $-40°C$ to $225°C$ temperature range was obtained based on the ZTC and "g_m/I_D" methodologies [4] and on the corresponding constant

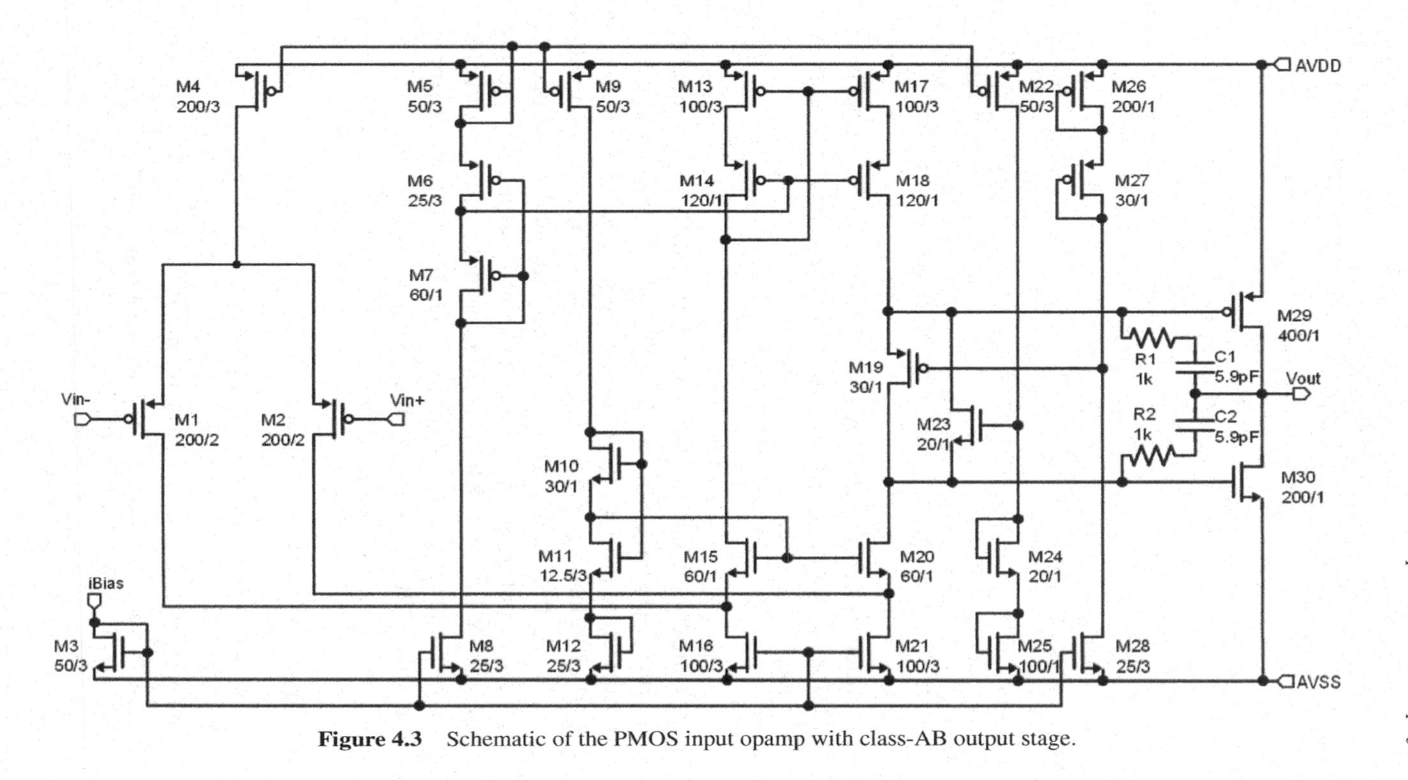

Figure 4.3 Schematic of the PMOS input opamp with class-AB output stage.

Table 4.2 Corner simulation results of the PMOS input opamp with class AB output stage

Specification	Min	Nom	Max	Units
Supply Voltage	4.5	5	5.5	V
Bias Current	30	55	70	μA
Current Consumption	221	459	804	μA
Temperature	–60	27	+225	°C
Input Voltage	0.5	2.5	3	V
Load Current	–100	50	100	μA
Capacitive Load	0	1.5	3	pF
DC open-loop Gain	70	111	114	dB
3-dB BW in Voltage-follower Config.	3	5.88	20	MHz
Phase Margin	57	71.82	101	o
Gain Margin	9	13.19	37	dB
Systematic offset voltage	–2.4	0.15	1.2	mV

bias current. Frequency compensation is obtained using *Miller* capacitors C_1 and C_2. The *Miller* capacitors determine the unity-gain frequency ω_0 of the opamp as given by:

$$\omega_0 = \frac{g_{m2}}{C_{Miller}}, \tag{4.1}$$

where g_{m2} is the transconductance of the input stage and C_{Miller} is equal to the C_1 and C_2 capacitors. Due to constant operating point, the transconductance reduces with temperature proportionally for all transistors, such that the poles and zeros keep their relative position. R_1 and R_2 have been chosen with the lowest temperature coefficient to ensure frequency stability over temperature. All devices were laid out as interdigitated devices to minimize the impact of thermal gradients. While the PMOS opamp was not measured directly, its performances were indirectly evaluated in the signal conditioning IA and in the SEDC. The signal conditioning channel and the SEDC were measured, and the results will be shown in Section 4.6.4.

4.1.3 NMOS-input OpAmp

An NMOS input class-AB operational amplifier was designed based on [2, 5, 6]. The schematic of the *OP*2 and *OP*3 operational amplifiers from Figure 4.58 is presented in Figure 4.4. The minimum input voltage should not be below $Vin_{min} = V_{tn1} + V_{ov1} + V_{ov4}$ volts, where V_{tn1} is the threshold voltage of M_1 and V_{ov1} and V_{ov4} are the overdrive voltages of M_1 and M_4, respectively. Vin_{min} almost doubles with the change of temperature from 200°C down to -55°C. *OP*2 is driven by the source follower (M_5) comprising an additional diode (D_1) in the current branch to provide the

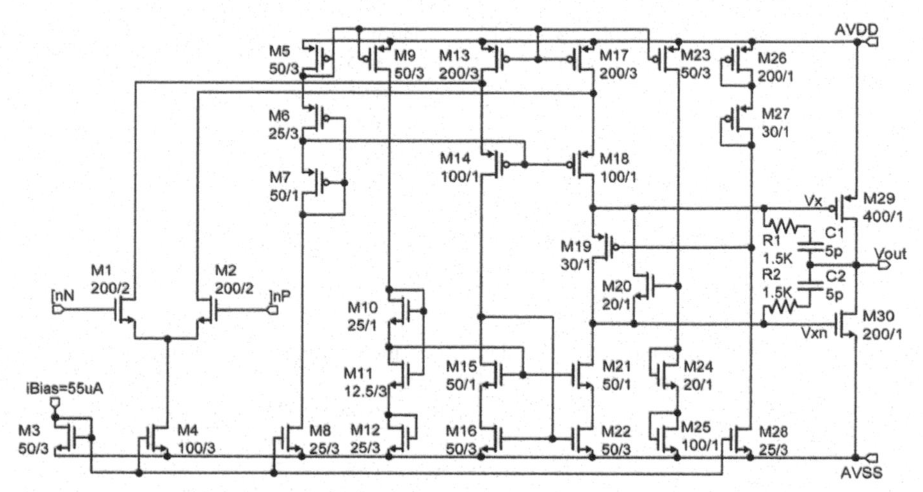

Figure 4.4 Circuit schematic, device sizes and bias current of the NMOS input class AB output stage opamp.

necessary minimum voltage biasing for the *OP*2 opamp, thus keeping all transistors of the input differential pair in saturation at all temperatures within the specified range. Cascode devices M_{15} and M_{21} are biased from the simplified Sooch current mirror $M_{10} - M_{11} - M_{12}$. Based on the ZTC and "g_m/I_D" methodologies [4] and on the corresponding constant bias current, a constant operating point over the −40°C to 225°C temperature range was achieved. This also guarantees that the transconductance reduces with temperature proportionally for all transistors, such that the poles and zeros keep their relative position. For a value of I_{dso} = 5.5 · 10^{-7} A, the input transistors M_1 and M_2 have been biased and sized for a value of g_m/I_D = 9.6 V^{-1} at 27°C and a value of 6.14 V^{-1} at 225°C, similar with values reported in [4]. The measured potential shift with temperature at the output of the class AB opamp connected in unity gain configuration is from 2.5 mV at 200°C to 6 mV at 250°C, similar with the results reported in [4]. The cascode voltage tracks the threshold and overdrive voltage change over temperature, thus properly biasing M_{15} and M_{21} at extreme temperatures. In order to guarantee the opamp stability over the specified temperature range, a bias current of i_{Bias} = 55 μA was used. The output class AB stage can provide currents of $\pm$100 μA, which is several times larger than the nominal quiescent current. The measured output impedance of Z_{out} = 586 Ω at 225°C shows that the Early voltage defining the output conductance at fixed bias current remains fairly constant.

Table 4.3 Process variation

CAP	RES	DIO	MOS
TM	TM	TM	TM
WP	WP	WP	WP
WS	WS	WS	WP
WP	WP	WP	WS
WS	WS	WS	WS
WP	WP	WP	WO
WP	WP	WP	WZ

Table 4.4 PVT corner simulation results of the NMOS input class AB output stage opamp. Process variation corners are presented in Table 4.3

Specification	Min	Nom	Max	Units
Supply	4.5	5	5.5	V
Bias Current	30	55	70	μA
Temperature	–60	27	+225	°C
Current Consumption	200	450	730	μA
Input Voltage	1.9	2.5	4.5	V
Load Current	–100	0	100	μA
Capacitive Load	0	1.5	3	pF
DC open-loop Gain	66	107	116	dB
Unity gain BW	2	8.4	18	MHz
Phase Margin	64	84.9	115	o
Gain Margin	10.9	22.11	35	dB
Systematic offset voltage	0.35	0.5	9.4	mV

Simulation results of the NMOS input opamp under a variety of process variation (Table 4.3), supply voltage and temperature (PVT) corners are shown in Table 4.4, where its robustness is evident. In Table 4.3, the acronyms TM, WP, WS, WO and WZ are denoting the process variation for typical mean, worst power, worst speed, worst one and worst zero, respectively. The three input opamp ($OP1$) from Figure 4.58 is used to control the charging PMOS device of the peak detector. The voltage at the input transistor M_{11} sets the minimum level for which the peak detector starts tracking the peaks of input signal. $OP1$ schematic is presented in Figure 4.5. With the exception of M_{11}, $OP1$ from Figure 4.58 has the same bias current, devices sizes and gain as $OP1$ from Figure 4.57.

4.2 Bandgap Reference Generator

The temperature stable bandgap reference voltage generator shown in Figure 4.6 is based on the work in [7]. It is a symmetrically matched

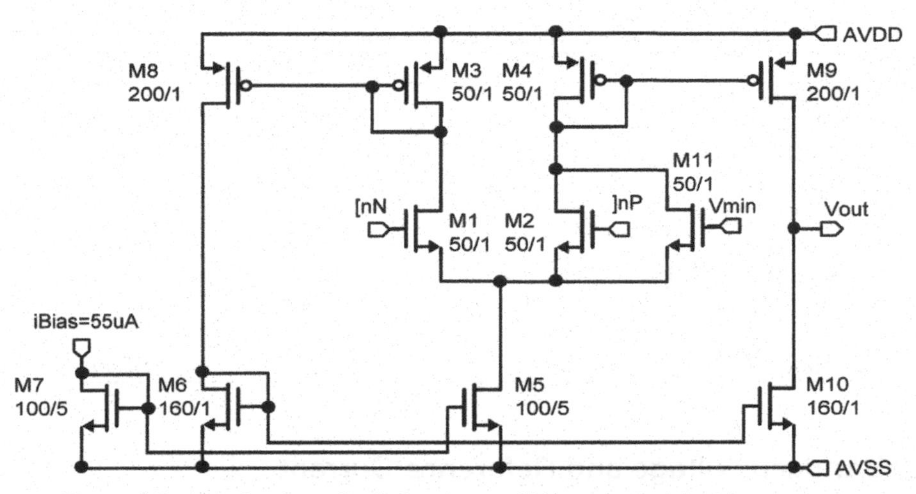

Figure 4.5 Circuit schematic, device sizes and bias current of the 3 input opamp.

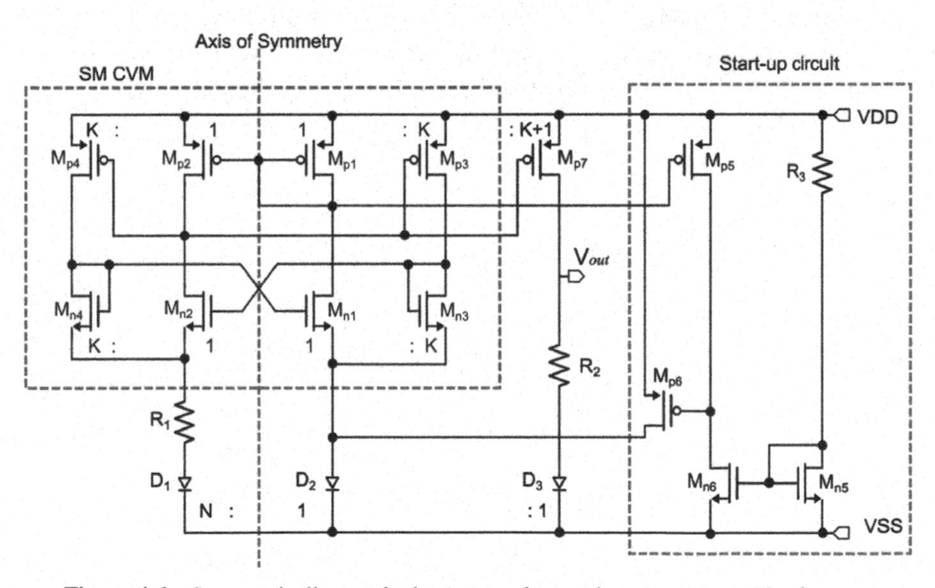

Figure 4.6 Symmetrically matched current-voltage mirror to generate V-reference.

current-voltage mirror. The start-up circuit is also shown. In the case where there is no current flowing in either of the D1 or the D2 branches, Mp6 will provide the start-up current until the drain of Mp6 has increased high enough (one diode voltage drop) to stop the injection of startup current. The typical simulated performance of the bandgap voltage reference is presented in Table 4.5.

Table 4.5 Bandgap voltage generator simulation results

Parameter	SMCVM	Unit
Supply Voltage VDD	5	V
Output Voltage V_{ref}	1.27	V
Supply Current	134	μA
Temperature Range	–60 to 260	°C
TC_{eff} at VDD = 3.3 V	24.4	ppm/°C
Line Regulation at 40°C	4.05	mV/V
Supply Sensitivity at 100 Hz	44	dB
Voltage Variation with temperature (per unit) $\frac{dV_{ref}}{dT}$	62.5	μV/°C
Noise [1 Hz to 100 MHz]	85.8	$\mu Vrms$

4.3 Bandgap Voltage and Reference Current

The bandgap voltage was measured across temperature and the results are shown in Figure 4.7. At an analogue supply voltage of 5 V, the bandgap voltage has a temperature coefficient of 151 ppm/°C between 25°C and 250°C, with a nominal voltage of 1.2041 V. The bandgap voltage drops by 3.4% between 25°C and 250°C.

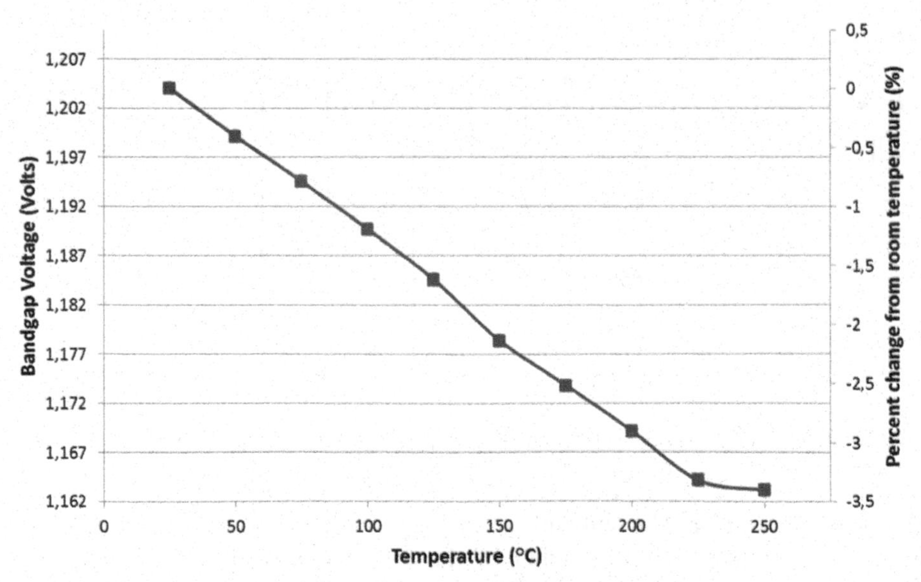

Figure 4.7 Bandgap voltage (actual and percent change) vs. temperature.

The reference current ($I_{reference}$) is generated by applying the bandgap voltage across an external 5.8 kΩ resistor with a low temperature-coefficient. At an analogue supply voltage of 5 V, the reference current has a temperature coefficient of 136 ppm/°C between 25°C and 250°C, with a nominal value of 201.7 μA. The reference current drops by 3.1% between 25°C and 250°C. Figure 4.8 demonstrates the layout of the bandgap voltage reference cell. The table in Figure 4.9 shows the nominal simulation results for the bandgap voltage reference cell including extracted post-layout parasitics.

4.4 Bias Network

Bias network generated reference voltage constant over temperature, reference current which is distributed across the chip and reference voltages for various analog cells of the ASIC. It consists of a bandgap voltage generator, a voltage-to-current converter, a voltage generator and current mirrors banks which replicate and distribute reference currents.

4.4.1 Top Level Schematic Diagram

The pin Vbg is connected to the bandgap output and routed to the pad to measure the bandgap voltage and overdrive it if necessary. Rport is a node for connecting an external resistor with low temperature coefficient for precise current generation. Iref_ovdrv pin is routed to the output pad. This is a backup pin which should be used to inject reference current in case of voltage-to-current converter failure. The rest pins are internal. IrefXX is denoting the reference bias currents with for OpAmps and comparators, while IexcXXX, IoXXX are denoting the high accuracy excitation and reference currents. vXXX is used for the reference voltages.

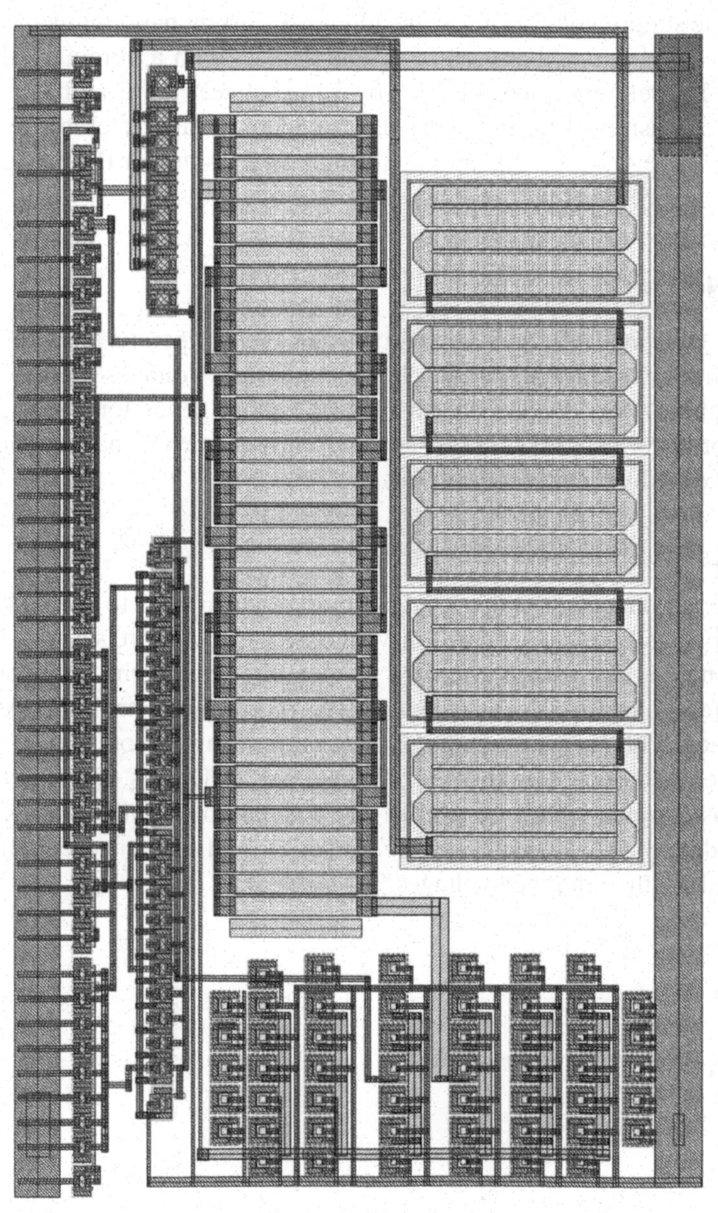

Figure 4.8 Layout of the bandgap voltage reference cell.

Parameter	SMCVM	Unit
Supply Voltage VDD	5	V
Output Voltage V_{ref}	1.27	V
Supply Current	134	μA
Temperature Range	-60 to 260	°C
TC_{eff} at VDD=3.3V	24.4	ppm/°C
Line Regulation at 40°C	4.05	mV/V
Supply Sensitivity at 100 Hz	44	dB
Voltage Variation with temperature (per unit) $\frac{dV_{ref}}{dT}$	62.5	μV/°C
Noise [1 Hz to 100 MHz]	85.8	μV_{rms}

Figure 4.9 Post-layout extraction simulation results of the bandgap voltage cell over PVT corners.

Table 4.6 Simulation results

Specification	Min	Nom	Max	Units	Note
Supply voltage		5		V	
Bias current	30	55	70	μA	
Current consumption			860	μA	
Temperature	–60		+225	°C	
Input voltage	0.5		4.5	V	
Load current	–100		100	μA	
Capacitive load	0		3	pF	
Input capacitance			0.8	pF	at 10 Hz
DC open-loop gain	64		113	dB	
3-dB bandwidth in voltage follower configuration	2.6		23.3	MHz	
Phase margin	52		99	o	
Gain margin	5.8		40	dB	
Integrated output noise in voltage follower configuration			0.15	mV	– Flicker noise is not included into simulation models – Ideal reference current source – Range: 10 Hz–1 GHz
Systematic offset voltage	–0.23		6	mV	Caused by output stage class AB biasing

4.4.2 Bias Network Layout

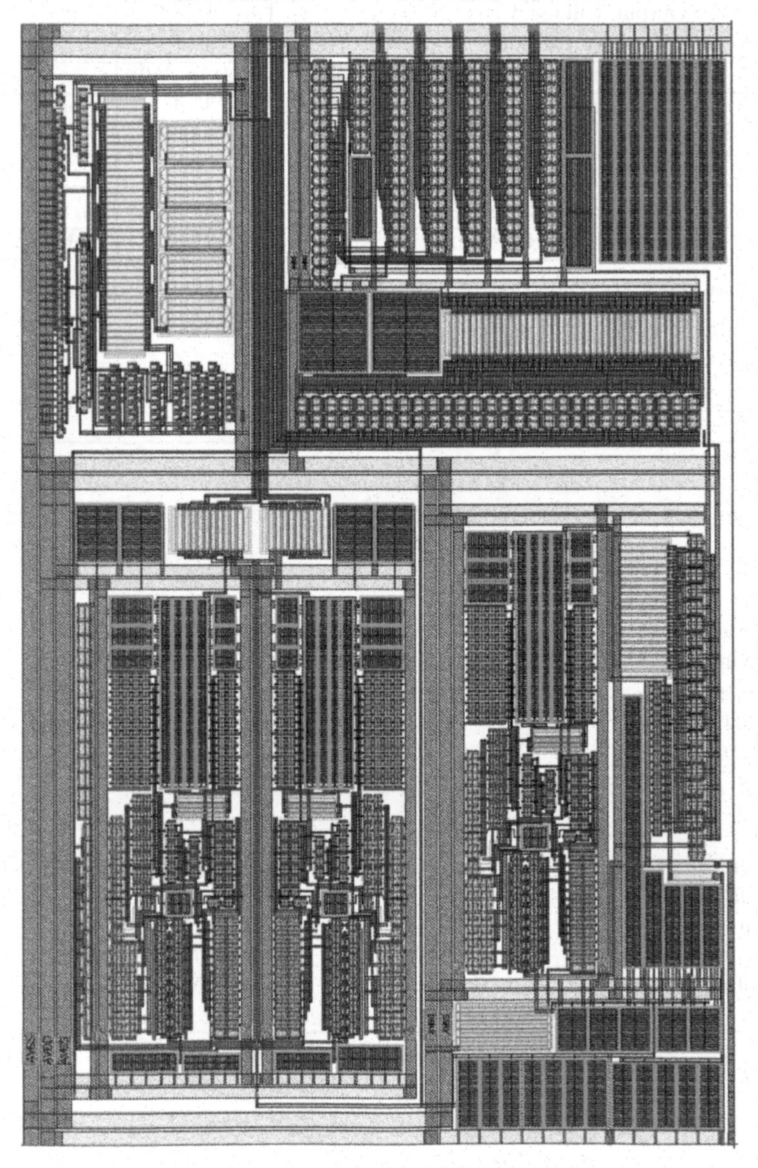

Figure 4.10 Bias network layout.

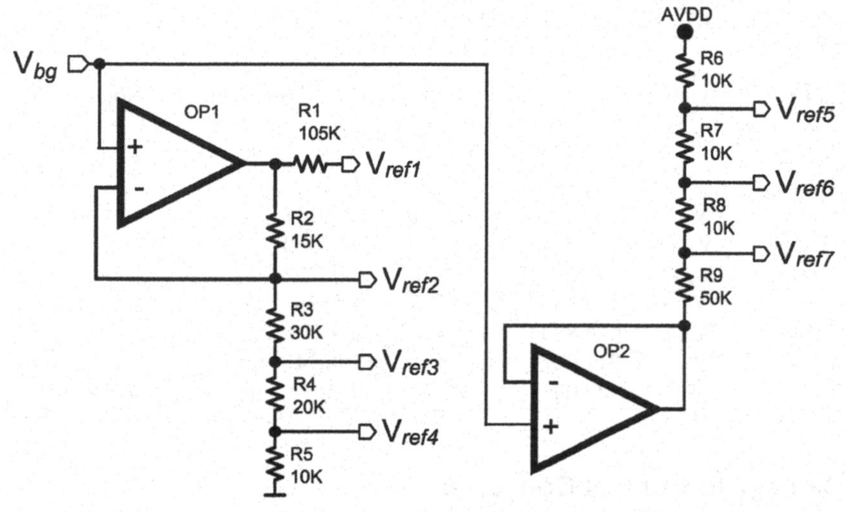

Figure 4.11 Reference voltages generator schematic diagram.

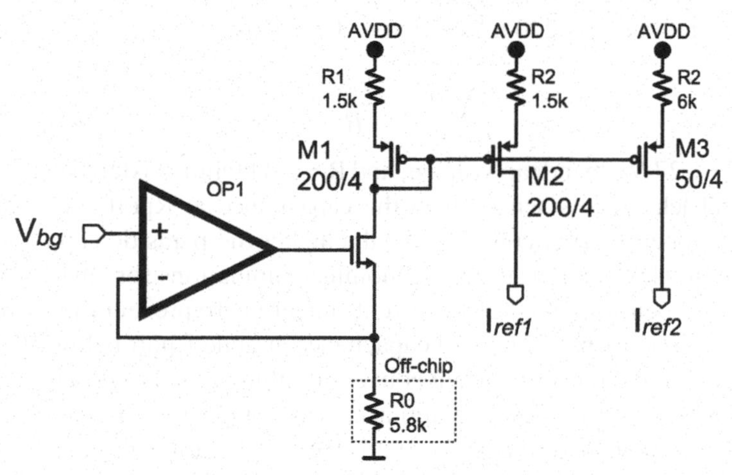

Figure 4.12 Schematic diagram of voltage to current converter.

4.4.3 Reference Voltages Generator

Reference voltage generator generates 6 different voltages from a bandgap output voltage. The values of the voltage values are: 3.6 V, 4.53 V, 4.06 V, 1.58 V, 634.9 mV and 211.6 mV. The simulation results of the voltage reference generator are presented in the Table 4.7.

Table 4.7 Simulation results of the voltage reference generator

		Value				
Specification	Symbol	Min	Nom	Max	Units	Notes
Temperature range		−50		250	°C	
Supply voltage			5		V	
Supply current		1.12	1.14	1.17	mA	
Output voltage	V_{3v6}	3.60145	3.601	3.60117	V	
Output voltage	V_{4v52}	4.53372	4.534	4.53382	V	
Output voltage	V_{4v05}	4.06745	4.067	4.06762	V	
Output voltage	V_{1v5}	1.5872	1.587	1.5876	V	
Output voltage	V_{0v6}	634.8834	634.9	635.050	mV	
Output voltage	V_{0v2}	211.6277	211.6	211.6514	mV	
RMS noise	V_{onRMS}	32.03	37.27	49.12	μV	Range: 1 Hz–1 GHz

4.4.4 Voltage to Current Converter

The voltage to current converter generated reference current using a stable bandgap voltage and a low TC external resistor. The nominal output current at Rport and Iref4x pin is 220 uA, which is defined as

$$I_{Rport} = \frac{V_{bg}}{R}$$

where Vbg = 1.27 V – bandgap voltage, and R = 5.8 kOhm – external resistor.

The feedback loop is sensitive to the capacitance at Rport. The phase margin of the open-loop circuit is lower the higher the parasitic cap at node Rport is. Figure 4.13 demonstrates the nominal simulations for small-signal stability of the open-loop voltage to current converter. To prevent ringing it is recommended to keep parasitic load capacitance at node Rport below 10 pF.

In the case of a voltage to current converter failure there is a possibility to define the reference current by an external current source. One of the possible failures of the voltage to current converter is the startup failure of the bandgap voltage reference. To disable the feedback loop pin Rport should be kept open or tied to AVDD via 1 MOhm external resistor. The external current is applied to the pin Iref_ovdrv. Another way to define the reference current externally is to apply the voltage at node Iref_ovdrv.

4.4.5 Current Mirrors

Current mirrors distribute generated reference currents across the ASIC. The cell consists of two mirror banks: the upper one with no source degeneration resistors for OpAmp and comparator bias currents (low precision current

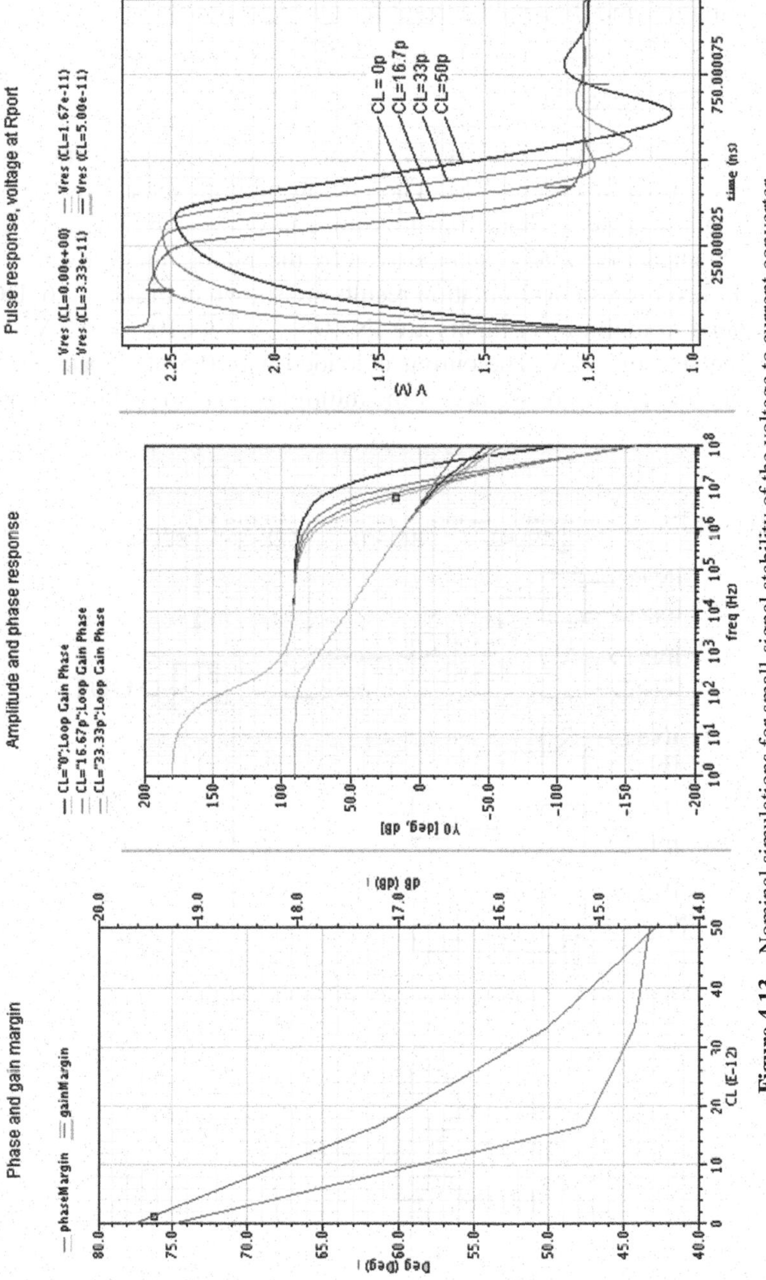

Figure 4.13 Nominal simulations for small-signal stability of the voltage to current converter.

mirror bank), and lower one with high accuracy current mirrors for excitation currents distribution and threshold current for QFreq channel.

4.5 Analog Multiplexer

The 11:1 analog multiplexer commutes 11 channels and test signals into one signal path as shown in Figure 4.14. The analog multiplexer consists of a series of transmission gates. Since transmission gates have some finite series resistance only capacitive loads can be driven by the analog multiplexer.

The multiplexer transient simulation results are shown in Figure 4.16. DC voltages applied to multiplexer inputs are IN<0> ... IN<10> = 0.2 V ... 4.2 V with the step of 0.4 V. The output is loaded with 0.5 pF capacitance. The highest nominal series resistance of the multiplexer is 8.6 kOhm at room temperature.

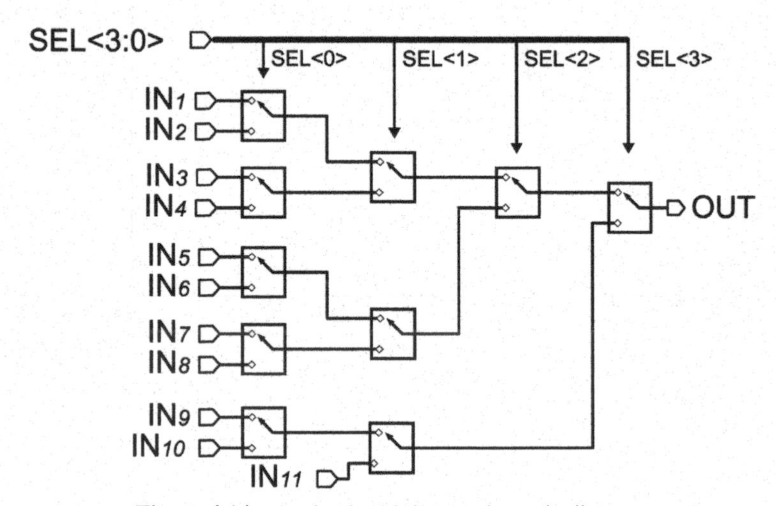

Figure 4.14 Analog multiplexer schematic diagram.

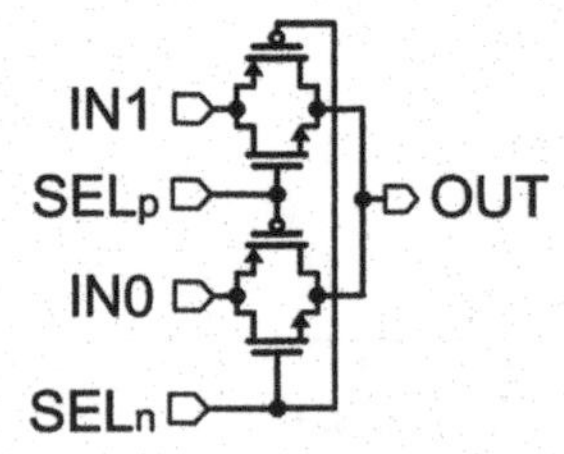

Figure 4.15 2:1 Multiplexer and transmission gate implementation.

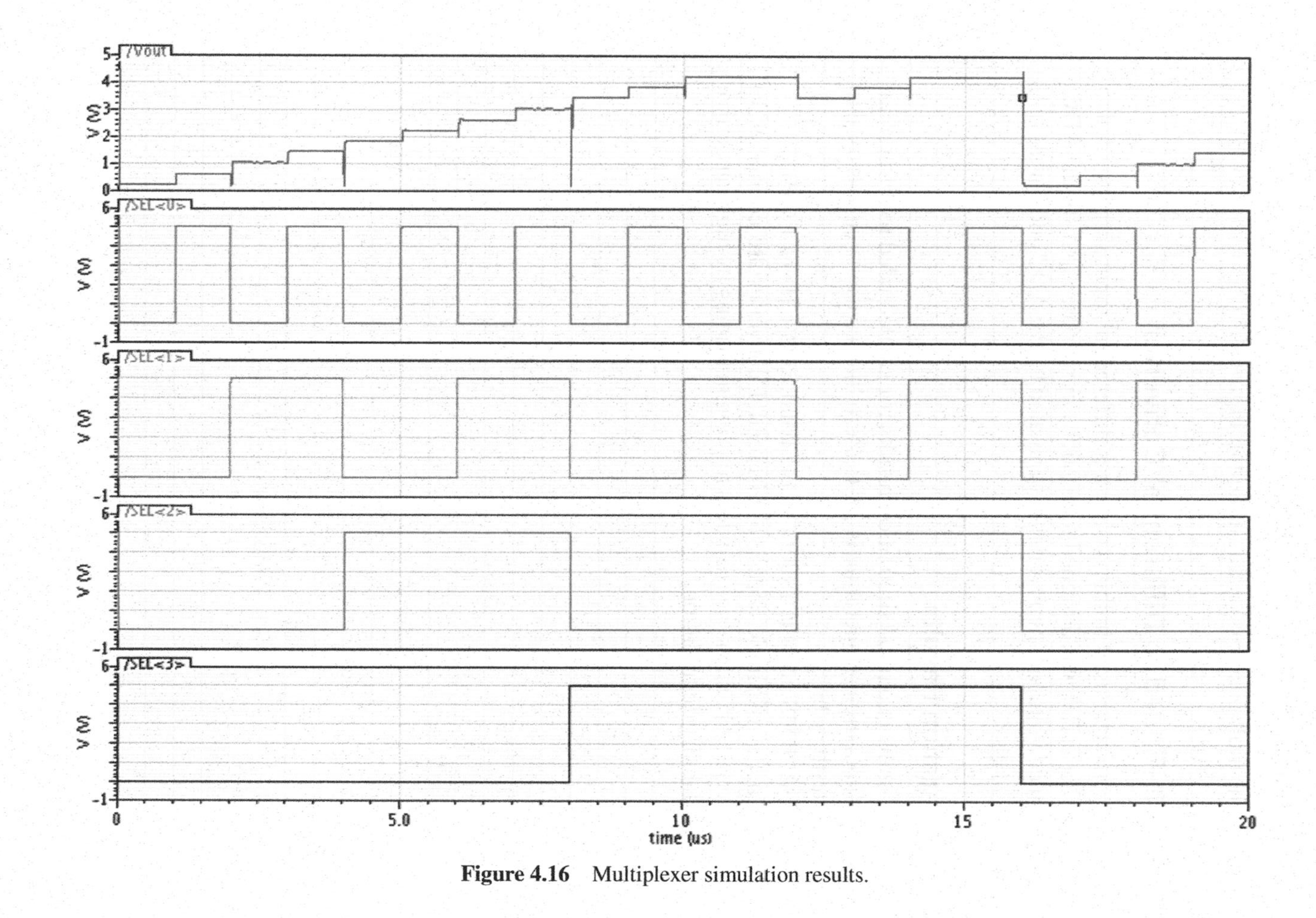

Figure 4.16 Multiplexer simulation results.

4.5.1 Layout of Analog Multiplexer

Layout of the multiplexer is demonstrated in Figure 4.17.

4.6 Single-Ended to Differential Converter

Since A/D converter used in this design has differential input single-ended to differential converter was designed to provide appropriate input for the ADC. Common mode voltage for converter is generated by the ADC and equals half of supply voltage.

4.6.1 Simulation Results

Figure 4.18 shows nominal simulation results for single-ended to differential converter. Simulation temperature is 27°C. Zero load capacitance was used in simulation testbench.

The output RMS noise is 190 uV in the band form 1 Hz to 1 GHz.

Output nodes of single-ended to differential converter are connected to output pads via 12 kOhm series resistors to prevent loading by high capacitive loads and stability degradation.

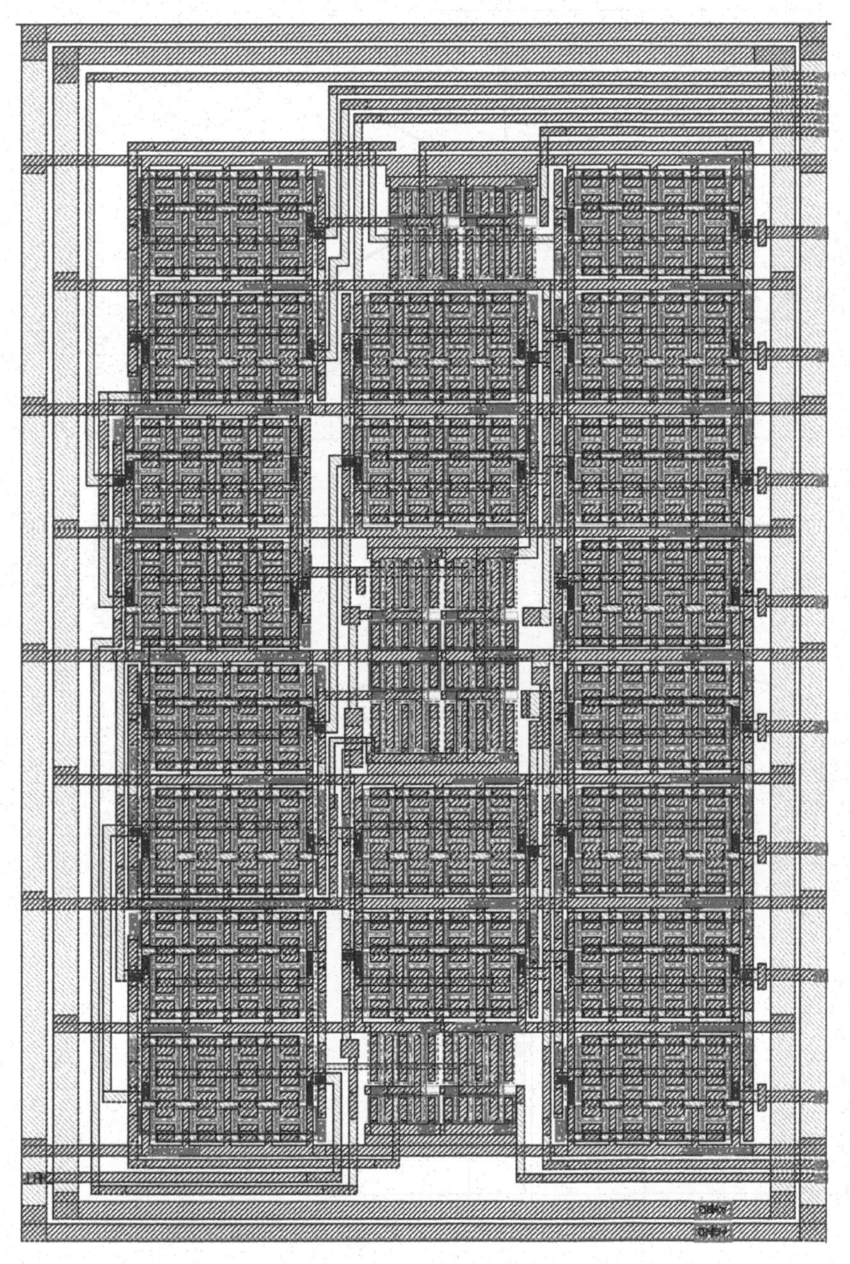

Figure 4.17 Layout of analog 11:1 multiplexer.

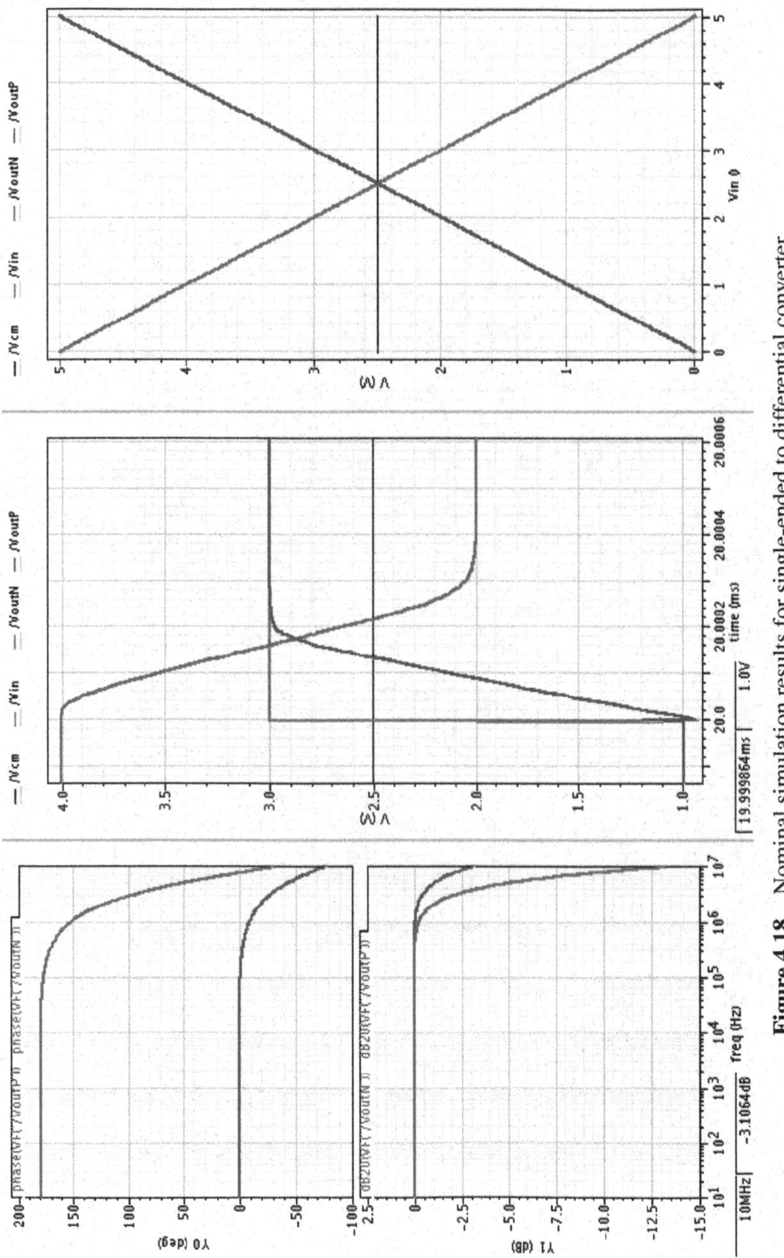

Figure 4.18 Nominal simulation results for single-ended to differential converter.

4.6.2 Layout of Single-Ended to Differential Converter

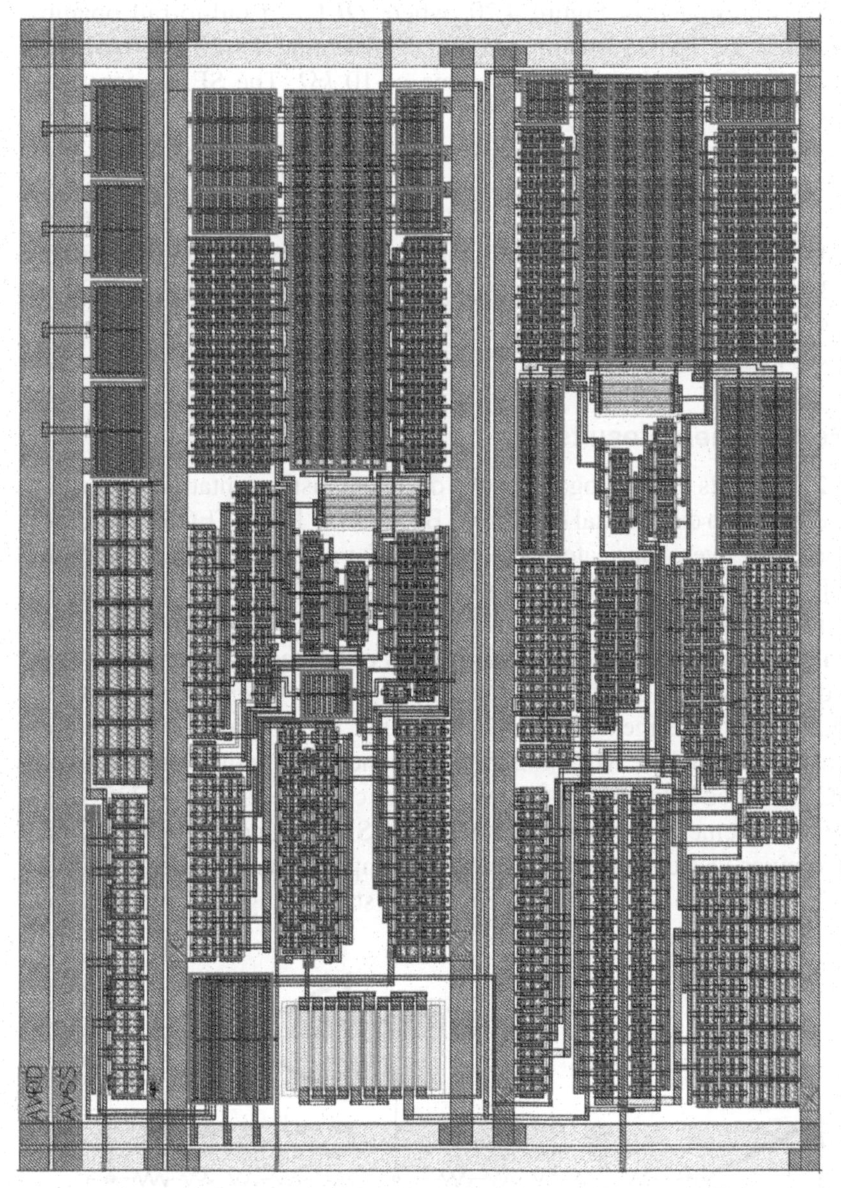

Figure 4.19 Single-ended to differential converter layout.

4.6.3 Single-Ended to Differential Converter

The SEDC is presented in Figure 4.20, where *OP*1 is a rail-to-rail opamp [3] and *OP*2 is the PMOS opamp presented in Section 4.1.2 [8]. The high resistivity $R1$ – $R10$ resistors have a value of 10 $k\Omega$. The SEDC converts a high or low impedance, single-ended input signal to a low impedance, balanced, differential output suitable for driving the following ADC. The reference common mode voltage is set at the V_{cm} input. V_{outP} is directly provided by *OP*1, while V_{outN} is the 180° phase shifted version of the input generated by *OP*2 and the $R1-R10$ resistors. On a single 5 V supply, the SEDC draws 1.2 mA, while the outputs can swing from 20 mV to 4.97 V. The SEDC can support SNR of 120 dB in a 1 kHz bandwidth and has a compact size of 750 μm × 540 μm.

4.6.4 Measurement Results

Figure 4.21 presents the micrograph of the designed instrumentation amplifier and single-ended to differential converter. Besides the HIGHTECS ASIC, the IA and the SEDC were fabricated on a test chip for electrical characterization over the [25–225]°C temperature range. The measured results of the temperature channel were presented in [1]. The measured DC gain of the IA used in the strain gauge signal conditioning channel is presented in Figure 4.22 and shows a difference of less than 1 dB over the [25–275]°C temperature range. Figure 4.23 shows the measured linearity of the PT100 based temperature channel at the input of the ADC at 225°C. The measured transfer function of the SEDC presented in Figure 4.24 shows a linear characteristic up to 225°C. A complete measurement system of the HIGHTECS ASIC, including the digital ARINC receiver was designed in VHDL and implemented on a Spartan 3E FPGA. The measured output waveform of the strain gauge channel via the

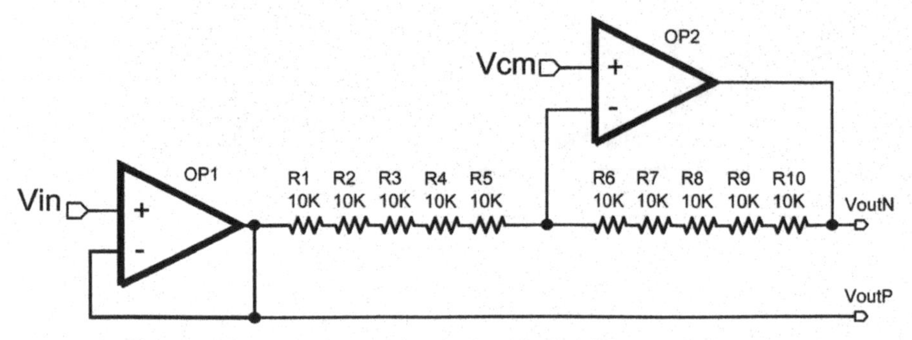

Figure 4.20 Schematic of the single-ended to differential converter.

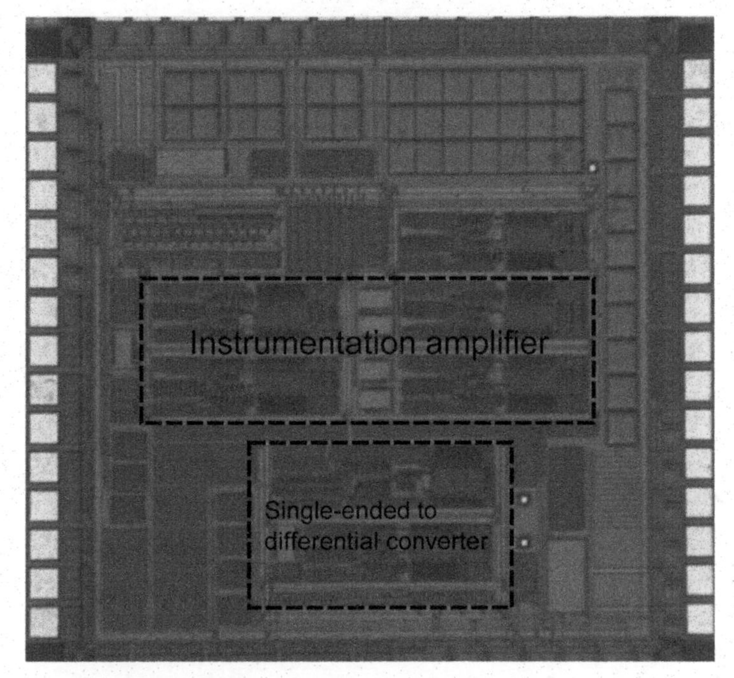

Figure 4.21 Micrograph of the designed instrumentation amplifier and single-ended to differential converter in X-FAB XI10 SOI process.

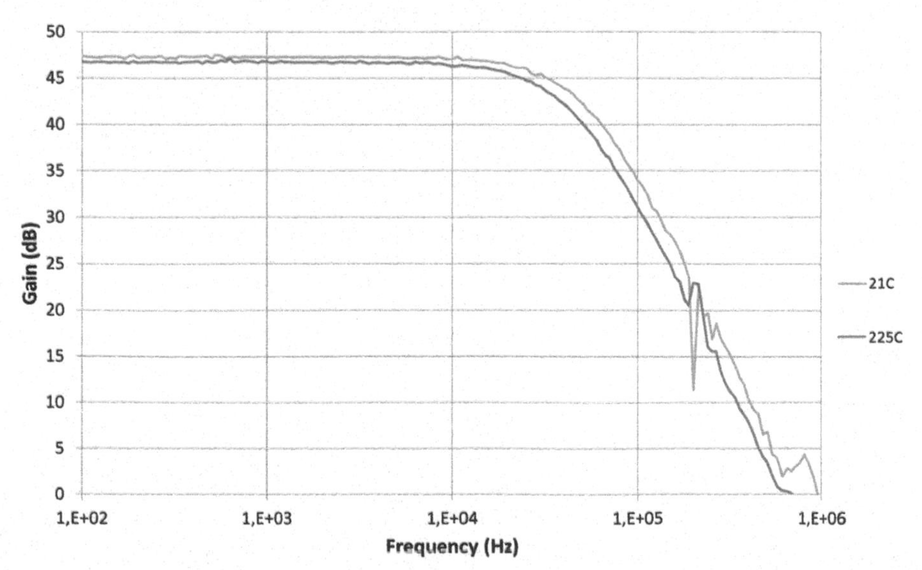

Figure 4.22 Measured DC gain of the instrumentation amplifier used in the strain gauge channel.

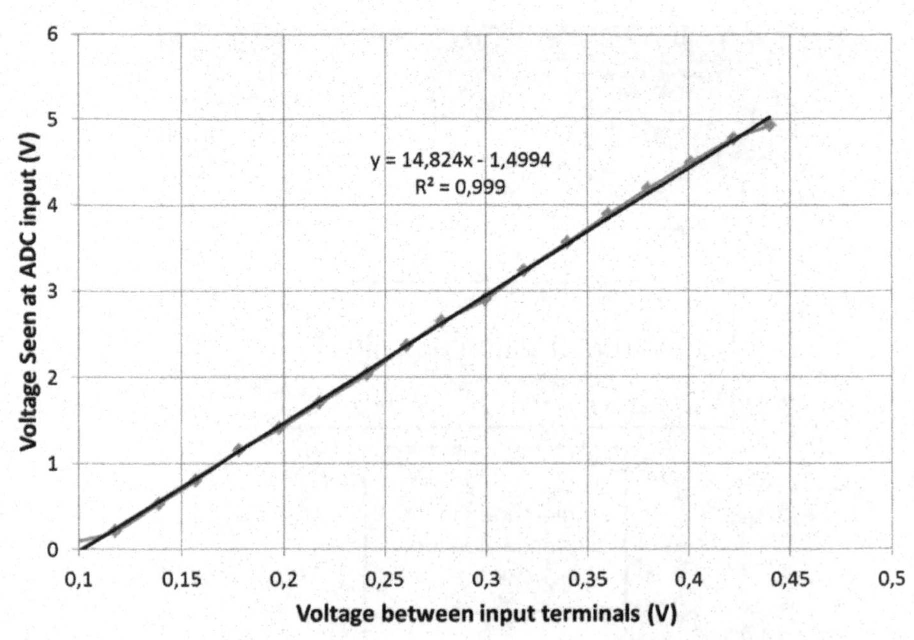

Figure 4.23 Measured linearity of the temperature channel at 225°C.

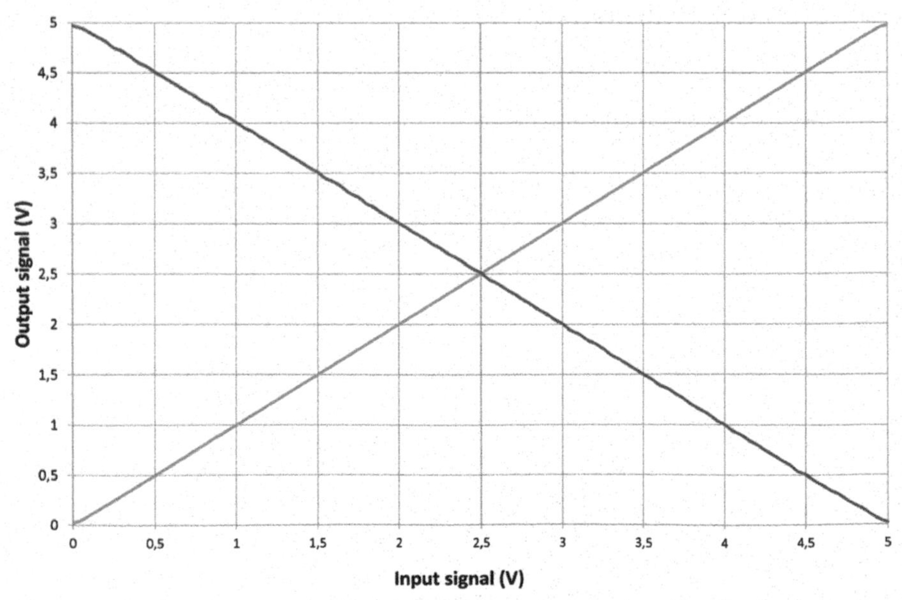

Figure 4.24 Measured transfer function of the single-ended to differential converter at 225°C.

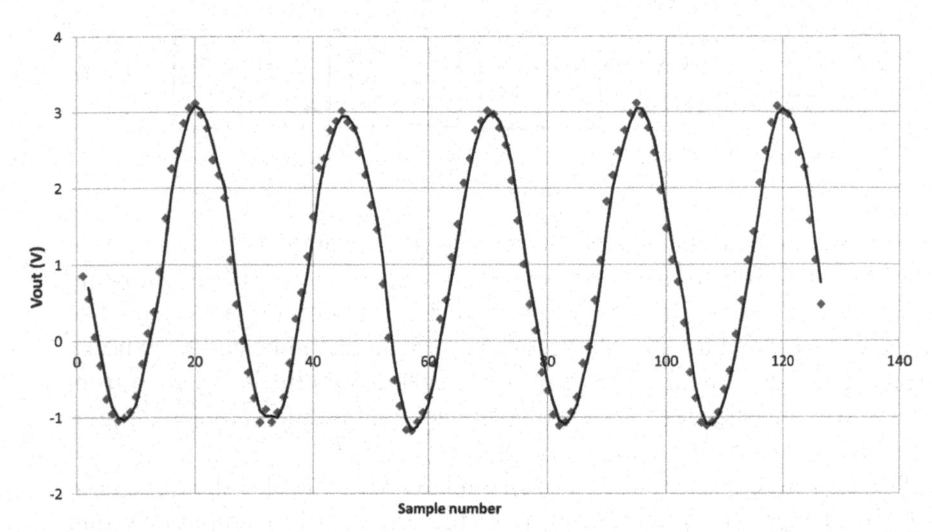

Figure 4.25 Measured output waveform of the strain gauge channel with a 16 mV sinusoidal input indicates a gain of 240.

Table 4.8 Specification versus achieved performance

Parameter	Specification	Measured	Units
T_{max}	200	225	°C
AVDD	5	5	V
IA DC Gain	47	46.8	dB
SEDC linearity range	[0.1–4.9]	[0.1–4.9]	V

IA, SEDC, ADC, ARINC transmitter and receiver is presented in Figure 4.25. This demonstrates the functionality of the complete signal conditioning chain from the input of the IA to the ARINC digital output. Table 4.8 presents the specification versus achieved performance. The IA and the SEDC are fulfilling their requirements at 200°C.

4.7 T1/TFo — Temperature Channels

4.7.1 Temperature Channels

The temperature channels T1, TFo1, and TFo2 are designed to excite a PT100 temperature-dependent resistor by applying a constant excitation current and conditioning the voltage signal coming off of the changing resistance. The block diagram for the sensor reading and signal conditioning is shown in Figure 4.26.

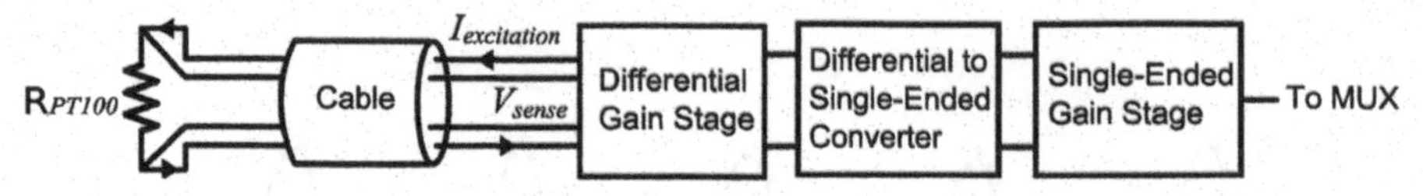

Figure 4.26 Temperature channel T1 signal conditioning diagram.

The excitation current is generated from a mirror of the reference current in the bias block. Ideally, this current does not vary over temperature, allowing the only signal variance to be the sensor resistance across temperature. The designed signal conditioning profile, where the blue area indicates the extreme voltages corresponding to a range of the excitation current ($I_{excitation}$) from 2.5 to 3 mA, and the green area shows the nominal profile corresponding to an excitation current of 2.7 mA.

The temperature sensor for these channels can be placed in the same or in a different temperature environment from the ASIC, as the excitation current will be minimally impacted by the temperature seen by the sensor. This channel also has been designed to determine and report open and short circuit fault conditions to the processing computer.

The excitation current measured in the previous section was applied across a PT100 RTD (Resistance Temperature Detector) at the TFo2 terminals. The measurement results are presented in Figure 4.27. The sensor was kept in the

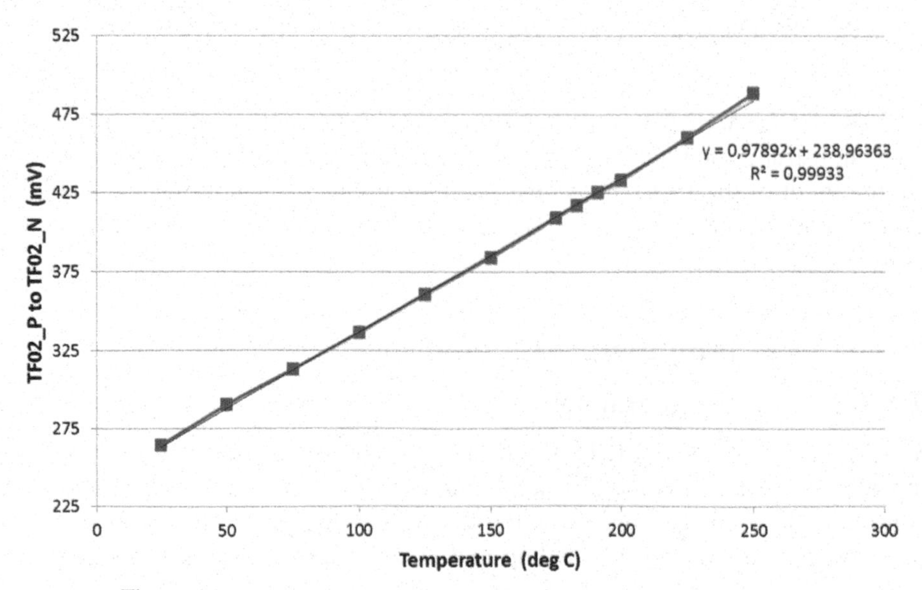

Figure 4.27 Voltage measured across TFo2 terminals vs. temperature.

same temperature environment as the ASIC and the resulting voltages were measured across temperature.

The RTD sensor is designed to be linear across temperature, and with an excitation current that has a low temperature dependence, the resultant voltage is highly linear across temperature. The data from a 5 volt analogue supply shows greater than 0.1% linearity (R^2 = 0.99933) between 25°C and 250°C.

PT100 is used as a sensor for T1 channel.

Sensor transfer function:

$$R_{PT100} = 99.4 + 0.379 \cdot T$$

Sensor gain:

$$k_{PT100} = 0.379 \frac{\Omega}{^\circ C}$$

For the specified temperature range

$$R_{PT100.\mathrm{min}} = R_{PT100}(-60^\circ C) = 77.66\Omega$$

$$R_{PT100.\mathrm{max}} = R_{PT100}(-130^\circ C) = 148.67\Omega$$

Taking sensor tolerances into account:

$$R_{PT100.\mathrm{min}} \geq 75\Omega$$

$$R_{PT100.\mathrm{min}} \geq 150\Omega$$

4.7.2 T1/TFo Operating Principle

Channel T1 comprises an instrumentation amplifier and an expiation current source which provides current for PT100 resistor. The voltage drop at PT100 is sensed by the instrumentation amplifier. Channel T1 uses four-terminal impedance sensing input.

Channel TFo is based on T1. The only difference is that TFo uses only two terminals for applying excitation current and sensing the voltage.

4.7.3 T1/TFo Functionality

4.7.3.1 Voltage-gain profile

The channel front-end incorporates three gain stages. The voltage-gain profile is shown in Figure 4.28.

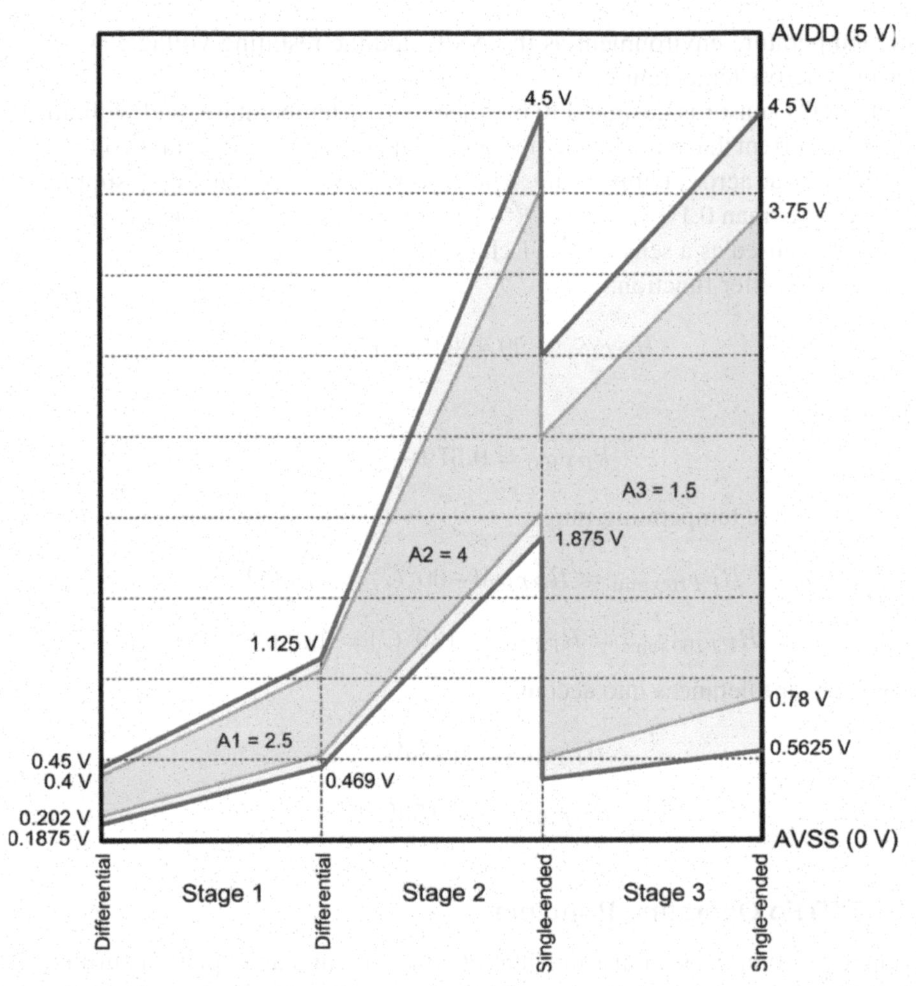

Figure 4.28 Voltage gain profile of the analog front-end. Blue area – extreme voltages, corresponding to 2.5 mA/3 mA excitation current. Green area – nominal profile, corresponding to 2.7 mA excitation current.

4.7.3.2 Static accuracy

$$\Delta T_{stat} = \frac{\Delta R_{stat}}{k_{PT100}}$$

where ΔT_{stat} is an equivalent resistance measurement error, defined as

$$\Delta R_{stat} = \Delta R_{IA} + \Delta R_{exc.temp} + \Delta R_{noise}$$

where ΔR_{IA} – resistance measurement error caused by an instrumentation amplifier,

$\Delta R_{exc.temp}$ – resistance measurement error caused by excitation current variation over temperature,

ΔR_{noise} – resistance measurement error caused by noise.

$$\Delta R_{IA} = \left(\frac{V_{out.meas} - V_{out.ideal}}{A_0} \right) / I_{exc.\min}$$

where $V_{out.meas}$ – measured voltage at the output of instrumentation amplifier,

$V_{out.ideal}$ – voltage at the output of an ideal instrumentation amplifier (no nonlinear distortions).

$$\Delta R_{exc.temp} = \frac{R_{PT100.\max} \cdot \Delta I_{exc.temp.\max}}{I_{exc.\min}},$$

Finally,

$$\Delta R_{noise} = \Delta T_{noise} \cdot k_{PT100} = \frac{V_{out.noise.\max}}{A_0 \cdot I_{exc.\min}}$$

4.7.3.2.1 *Static accuracy simulation for variable chip temperature and constant sensor temperature*

The sensor temperature is kept at 130°C, which is the maximum specified temperature. Model sections for simulation are changed simultaneously for all elements (all-wp, all-tm etc.). The static error is calculated using the output voltage deviation from a constant value over temperature, gain of the IA and the gain of the sensor:

$$\Delta_T = \frac{\Delta V_{out}}{A_0 \cdot I_{exc.\min} \cdot k_{PT100}} = \frac{\Delta V_{out}}{15 \cdot 2.5\ mA \cdot 0.379 \frac{\Omega}{°C}}$$

The accuracy is simulated with transistor-level bandgap and excitation current source and ideal 1.2 V reference and ideal excitation current source. Results shown here do not take into account instantaneous temperature error caused by the noise of instrumentation amplifiers. Simulation results of T1's output voltage over a variety of corner variation is presented in Figure 4.29. Transistor-level circuits for the bandgap voltage reference and excitation current source were used. The largest static error over three simulated cases is 4.5°C (±2.26°C) at –60 . . . +250°C temperature span, and 1.41°C (±0.7°C) at +50 . . . +150°C.

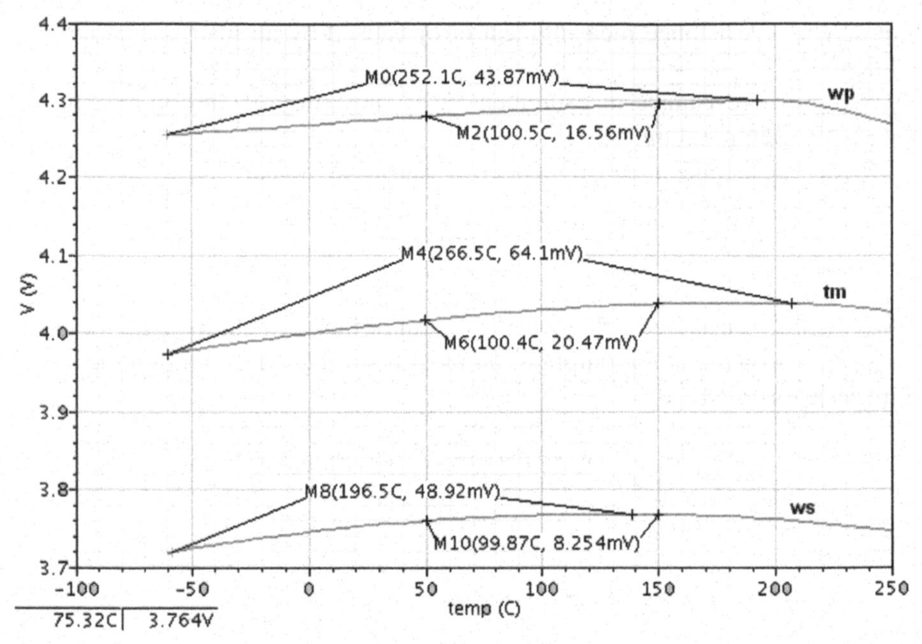

Figure 4.29 Output voltage of T1 channel. Transistor-level bandgap and excitation current source. The largest static error over three simulated cases is 4.5°C (±2.26°C) at –60 . . . +250°C temperature span, and 1.41°C (±0.7°C) at +50 . . . +150°C.

4.7.3.2.2 *Static accuracy simulation for variable chip temperature and variable sensor temperature*

The temperature range for this simulation is $-60°\text{C} \ldots +130°\text{C}$, as specified in specs. The chip temperature follows the sensor temperature.

4.7.3.3 Temperature error due to quantization error

Assuming 12-bit ADC and 4 V input voltage range, temperature error caused by quantization error is:

$$\Delta T_q = \frac{\Delta R_q}{k_{PT100}} = \frac{\Delta V_{out.q}}{A_0 \cdot k_{PT100} \cdot I_{exc.\min}} = \frac{4/2^{12}V}{15 \cdot 0.379\frac{\Omega}{°C} \cdot 2.5\ mA}$$
$$= 0.07°C$$

4.7.3.4 Input ESD protection

Voltage levels at chip inputs are kept below 5 V at normal operating conditions. A pad cell used for T1/TFo inputs is "APRBDF". The schematic of the ESD protection circuitry is shown in Figure 4.32.

4.7.3.5 Short and open circuit detection

The following short/open conditions are detected:

- Short at pins MP and MN;
- Open at pins MP and MN;
- Open at pins EP and MP, EN and MN (channel T1 only)

The following short/open conditions are not detected:

- Short at pins EP and MP, EN and MN (channel T1 only)

4.7.4 Mirrored Bias Current for Temperature Probe Excitation

Three of the temperature channels have been designed to work with temperature-dependent resistors like the PT100. To be able to measure the resistance, and thus compute the temperature seen by the sensor, an excitation current is provided via a multiplying mirror of the reference current.

Excitation current mirror contains source degenerating resistors to increase output impedance of the current sources and reduce current dependence on temperature. Simulation results of T1's output voltage over a variety of corner variation is presented in Figure 4.30. Ideal excitation current source

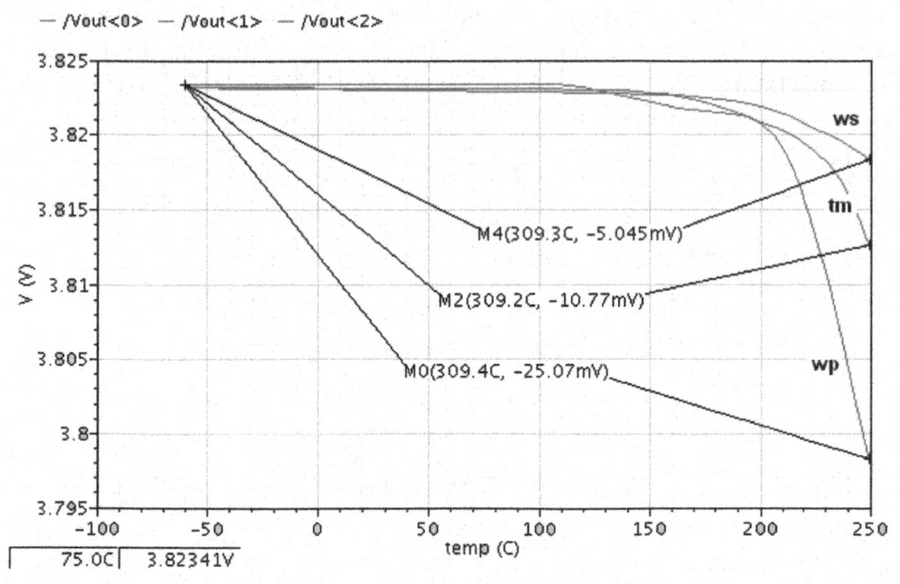

Figure 4.30 Output voltage of T1 channel. Ideal excitation current source and reference voltage source. The largest static error over three simulated cases is 1.76°C (±0.88°C) at –60 . . . +250°C.

and reference voltage source were used. The largest static error over three simulated cases is 1.76°C (±0.88°C) at –60 . . . +250°C.

The simulated resistance error over corner variation is presented in Figure 4.31. Transistor-level circuits for the bandgap voltage reference and excitation current source were used. The largest static error over three simulated cases is 0.67 Ohm, which corresponds to 1.77°C (±0.89°C) at –60 . . . +130°C.

4.7.5 T1/TFo Channels Schematic Diagrams

4.7.5.1 T1 top level connection

No off-chip discrete components for T1/TFo are required.

The top level schematic of T1 channel is presented in Figure 4.33.

Channel TFo2 was measured across temperature and the results are shown in Figure 4.34. The excitation current is given by the solid line and the mirroring ratio between the excitation current and the reference current is given by the dotted line.

The excitation current has a temperature coefficient of 169 ppm/°C between 25°C and 250°C, with a nominal value of 2.43 mA at room temperature. The mirroring ratio has a temperature coefficient of 326 ppm/°C between 25°C and 250°C, with a nominal designed ratio of 12.5. These low temperature coefficients are indicative of the stability of the biasing block across temperature.

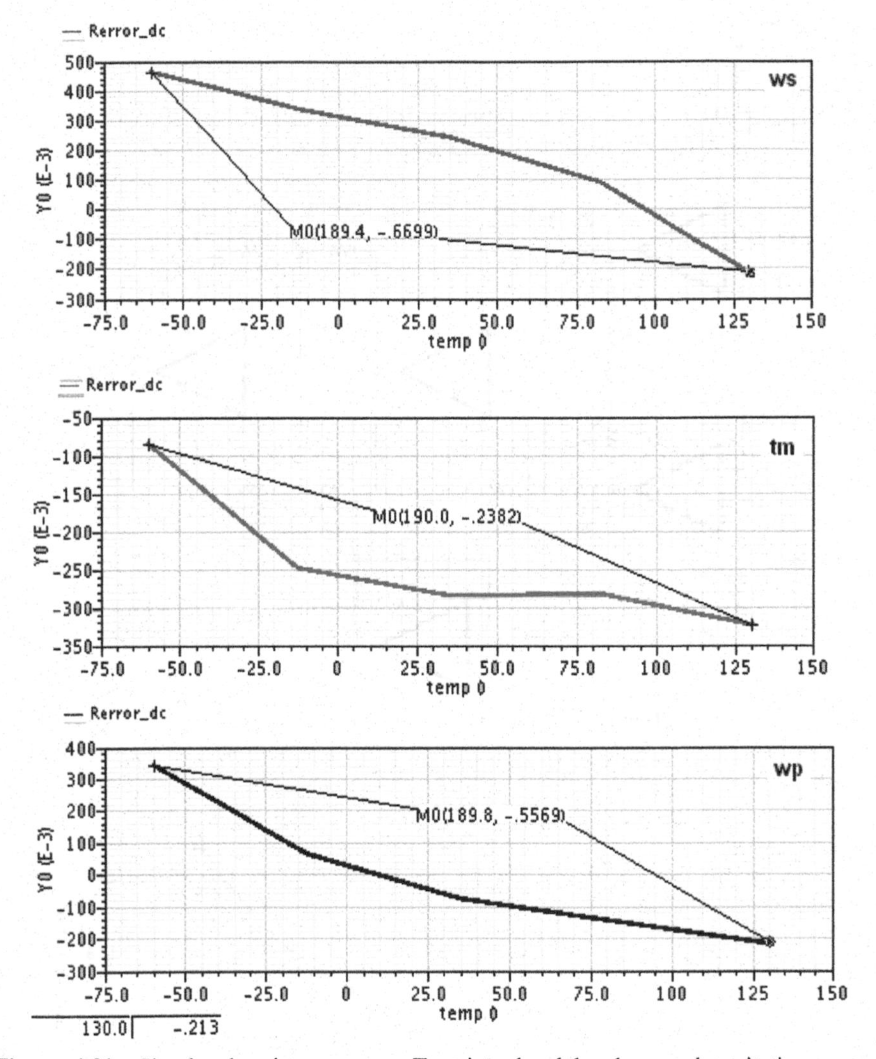

Figure 4.31 Simulated resistance error. Transistor-level bandgap and excitation current source. The largest static error over three simulated cases is 0.67 Ohm, which corresponds to 1.77°C (±0.89°C) at –60 . . . +130°C.

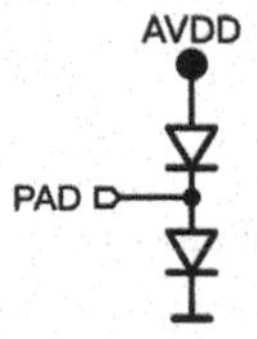

Figure 4.32 Input pad for channel T1/TFo (“APRBDF”).

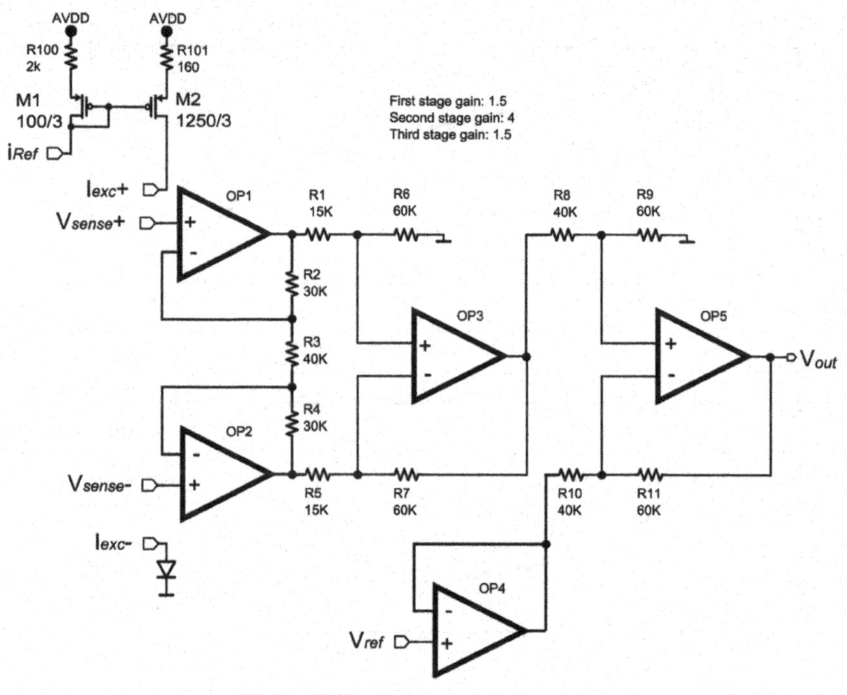

Figure 4.33 T1 top level schematic.

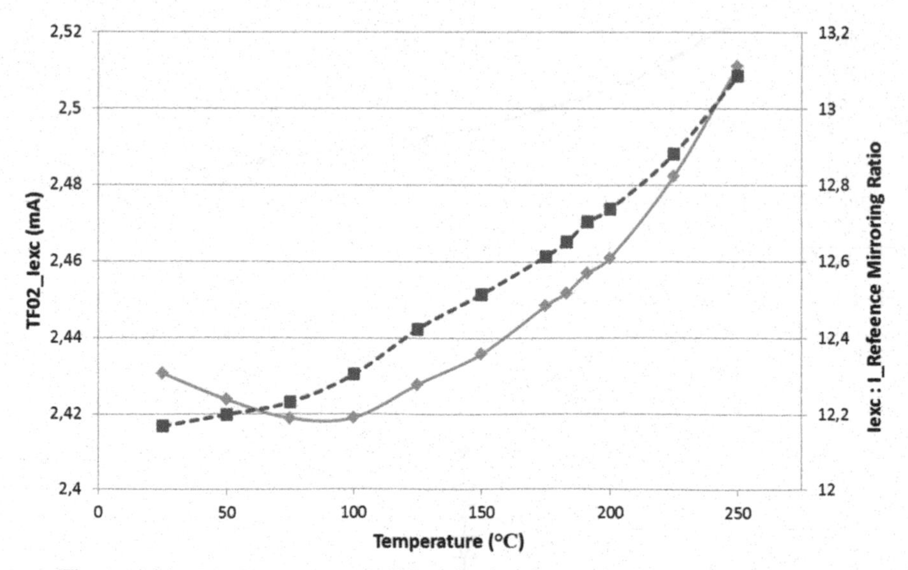

Figure 4.34 TFo2 $I_{excitation}$ (solid) and mirroring ratio (dotted) vs. temperature.

4.7.6 T1/TFo Channels Layout

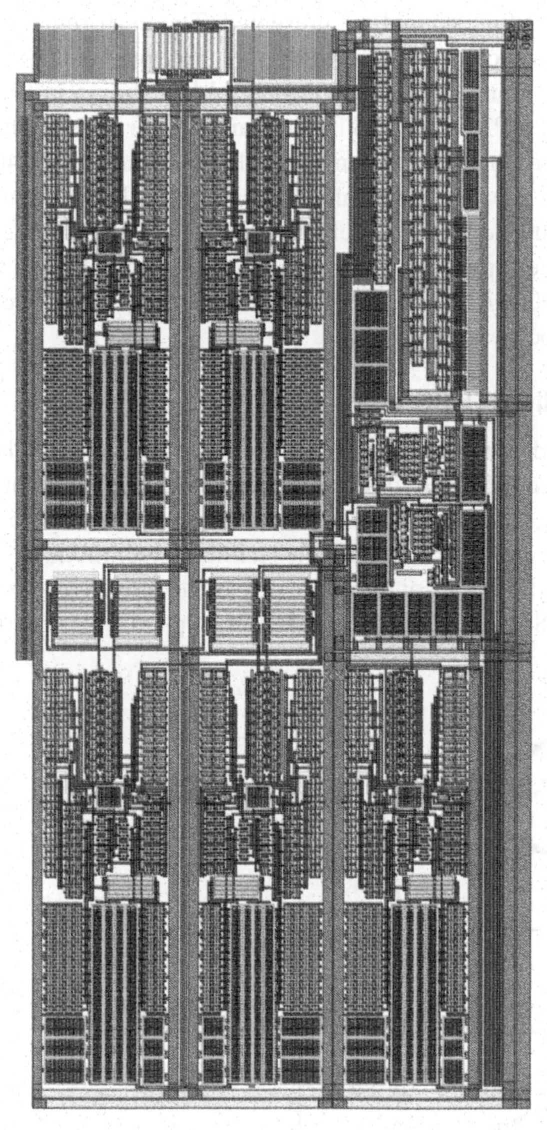

Figure 4.35 T1/TFo layout – size: 730 um × 1530 um.

4.8 SG2 — Strain Gauge Channel

In this section the error budget of the PT100 based temperature channel and the IA for the strain gauge channel are presented. The performance of the IA used in the temperature channel was previously presented in [1]. The block diagram for the strain gauge signal conditioning IA is shown in Figure 4.36, where *OP*1, *OP*2 and *OP*3 are PMOS input opamps. The IA is connected to a strain gauge bridge and has to perform a signal amplification of 240 (47 dB) based on the sensor's measurement range specification. In addition to signal amplification, the conditioning channel has to detect the *OPEN* or *SHORT* to 10 V conditions of the input terminals. The IA draws 3 mA from a 5 V suply and has a size of 1290 $\mu m \times 725\ \mu m$. The voltage at V_{ref} is provided by the voltage reference block presented in [1].

4.8.1 Testing of HIGHTECS Module

The HIGHTECS module was assembled as shown in Figure 4.37, the module consisted of the HIGHTECS hybrid circuit was selected to be built into a module to be tested, and the high temperature printed circuit board with resistors wired to the connectors within the stainless steel module.

The output from SG2 sensor was selected for testing over the range of voltage MN9.2 to 9.8 V with MP set at 9.5 V in steps of 0.05 V from –40°C to +225°C.

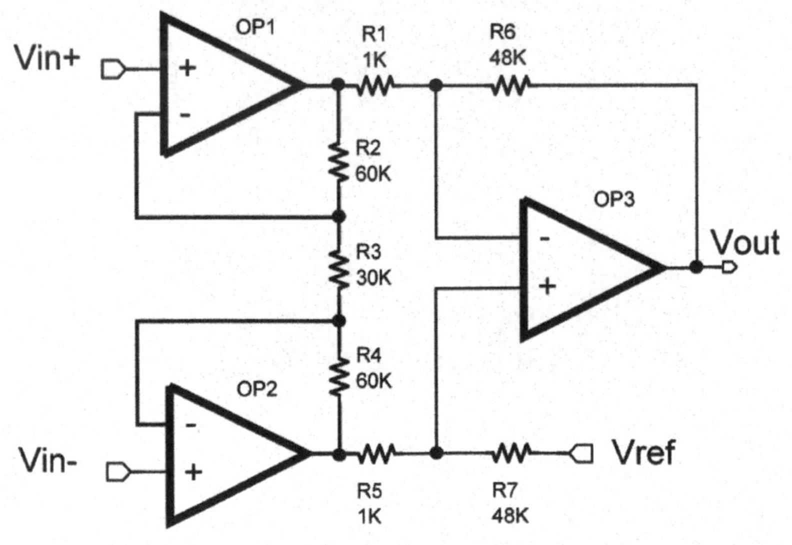

Figure 4.36 Simplified schematic of the strain gauge signal conditioning channel.

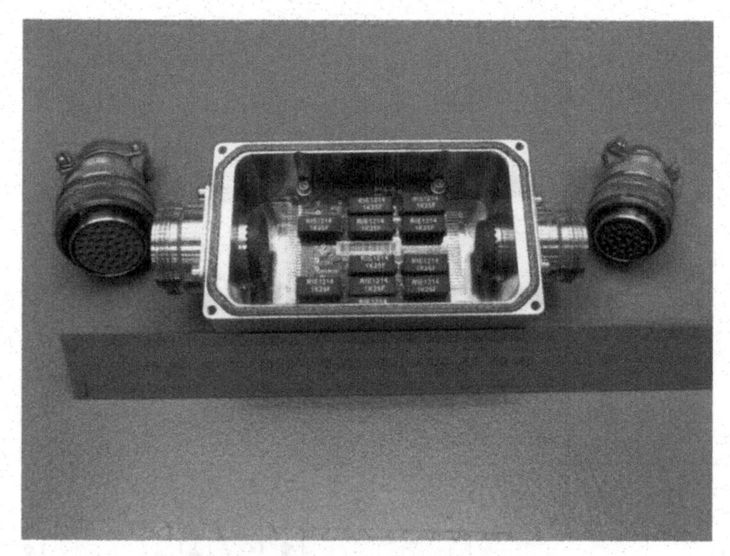

Figure 4.37 HIGHTECS module.

The results show a predominantly linear output at temperatures of 20°C and above, when the outputs with error codings are removed from the analysis.

4.9 QFREQ — Frequency Channel

4.9.1 Introduction

In this section, we focus on the mixed signal conditioning and digital counting units to calculate the frequency of a signal coming from an electromagnetic sensor facing a phonic wheel of an aeroengine [3, 9]. The phonic wheel consists of two intermeshed teeth on the loaded engine drive shaft and another set of teeth on the unloaded shaft end in the same plane. The produced signal enables to measure the frequency and the phase between the loaded and unloaded phonic wheel enables to calculate the torque. The sensor signal can have amplitudes up to 70 V_{pp} with sharp edges and a maximum frequency of 4 kHz.

A novel CMOS SOI current peak detector was developed to detect and process the input signal [3]. Current peak detectors have been previously published in [10–13]. In [10, 12], the disadvantage of the peak detectors is that the hold capacitor is charged by the input current. In Figure 4.38 [11], for a high sensitivity of the peak detector, the output resistance of the feedback transistor PI and the input current source I_{hr} must be high, while for a high

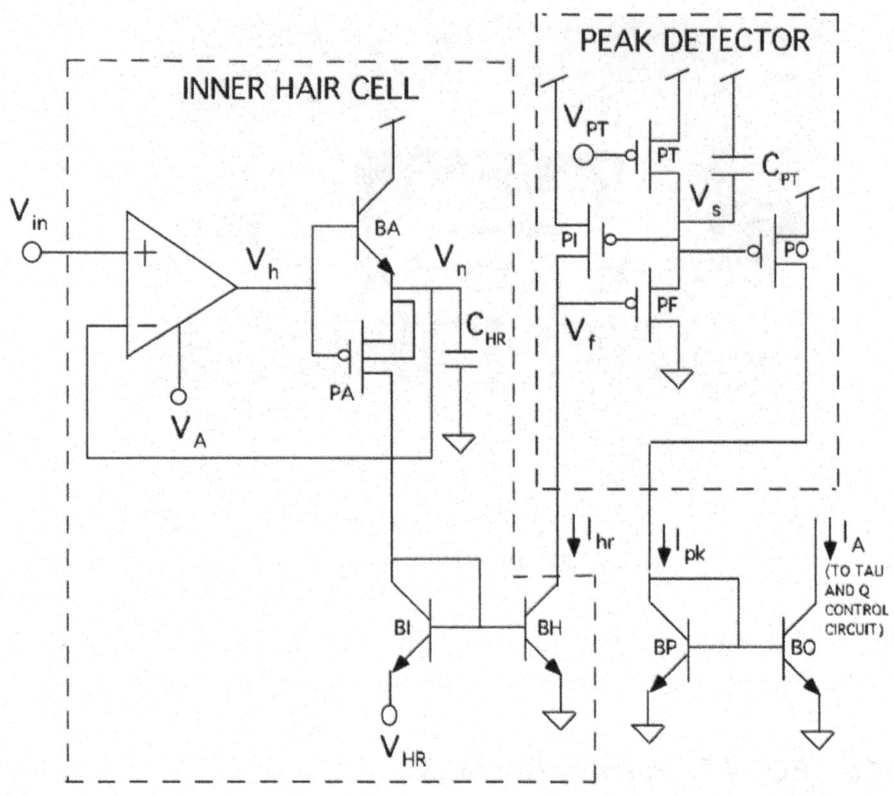

Figure 4.38 Peak detector presented in Figure 9 of [11]. (c) Springer. Reprinted with permission.

charging speed, the output impedance of the *PI* transistor and I_{hr} must be low to provide good driving strength for the *PF* transistor. In [13], Figure 4.39, the peak detector is a trade off between the sensitivity and speed, as follows: for a high sensitivity the output of M_2, M_3 and I_a must be high, while for a high charging speed, the output resistance of M_3 and I_a must be low to provide good driving strength for M_5.

The section is organized as follows. Section 4.9.2 describes the system architecture, the input signal definition and derives the unit specifications. Section 4.9.3 presents the operation principle and the design methodology of the proposed detector. The proposed CMOS SOI current peak detector is presented in Subsection 4.9.3.2. Its main advantages over previously published architectures are stable operation over a wide temperature range and high

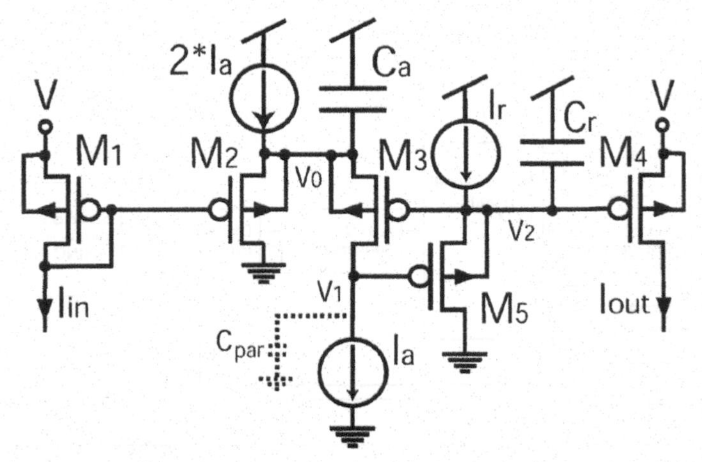

Figure 4.39 Peak detector presented in Figure 2 of [13] (c) IEEE. Reprinted with permission.

speed due to low charging time of the external capacitor. The proposed pulse selector is presented in Subsection 4.9.3.4. Their main advantages are that they are internally generating and processing digital pulses, therefore eliminating the need of an analog-to-digital converter (ADC) in the signal path. Experimental results and conclusions are presented in Sections 4.9.4.

4.9.2 System Architecture

The HIGHTECS ASIC is made up of several system level blocks shown in Table 4.9 and connected as in Figure 4.40.

Functionally, the reference voltage generator and bias circuits provide the needed voltages and currents for the opamp-based signal conditioning circuits for each type of sensor. The signal flow then goes from the excitation of the

Table 4.9 Functional blocks included in the ASIC

List of Functional Blocks	
Bandgap Reference Generator	4 Temperature Channels
Reference Current Generator	4 Strain Gauge Channels
Bias Voltages Generator	3 Pressure Sensor Channels
Global Current Mirrors	ADC
Digital Input	ARINC 429 Driver
Frequency Signal Conditioning	ARINC 429 Control Sequencer
Frequency Pulse Counting Logic	

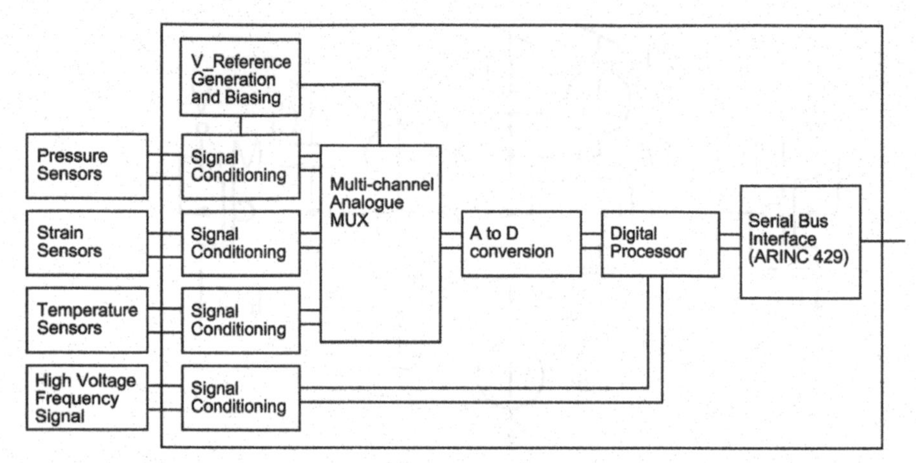

Figure 4.40 HIGHTECS ASIC function level block diagram including the signal conditioning processing the high voltage frequency signal.

sensor, to the signal conditioning, to the multi-channel analogue multiplexer (MUX) and the ADC, and then to the ARINC429 [14] databus and out to the processing computer. The serial bus interface (ARINC 429) facilitates the data transmission from the ASIC to the FADEC or EHMS. With the exception of the ADC, all of the other HIGHTECS ASIC blocks have been customly designed during the project. The signal conditioning unit processing the high voltage frequency signal was integrated onto the HIGHTECS ASIC [1, 3], and requires only two pairs of external resistors and capacitors for signal conditioning and six external resistors for ESD protection. The block diagram of the frequency signal conditioning unit is presented in Figure 4.41. The input stage includes the input diodes, the current mirrors and the ESD protection. The current path (Ipath) includes the current peak detector, the current divider and the current comparator. The voltage path (Vpath) includes the voltage-peak detector, the voltage divider and the voltage comparator. The input diodes perform the voltage-to-voltage or voltage-to-current conversions. The current mirrors are copying the current to the following stages in the current and voltage path. The ESD pads ensure that the voltage at the input of the ASIC does not exceed 5 V during normal operating conditions. The current peak detector includes a current comparator, an opamp-based voltage peak detector and a current source. The pulse selector is generating a digital output signal highly immune to harsh environment conditions. The frequency pulse counting logic is embedded into the digital processor following the HIGHTECS ADC.

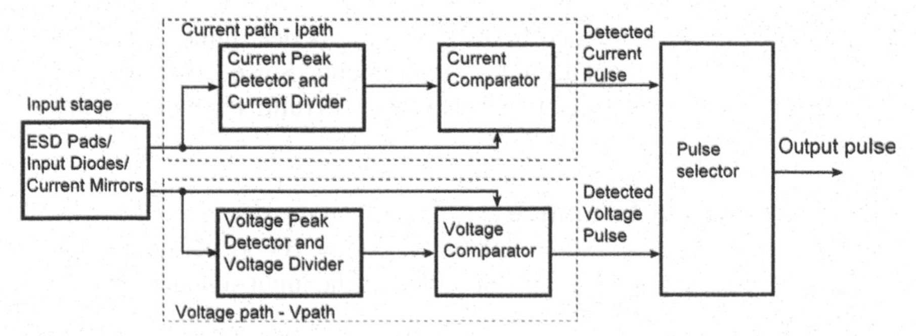

Figure 4.41 Block diagram of the frequency signal conditioning unit for rotating equipment.

The circuit implementation of the frequency signal conditioning unit is presented in Section 4.9.3.

4.9.2.1 Input signal definition

The definition of a typical input signal is presented in Figure 4.42, where A is the signal amplitude, T is the signal period, N_{max} is the maximum noise voltage, V_t is the threshold voltage, ΔV_t is the target hysteresis window opening, A is the average signal peak level, and ΔA_i is the instantaneous signal peak level deviation. The parameters of the rotational system are defined as F_{min} = 50 Hz and F_{max} = 4 kHz. The frequency signal conditioning unit has to provide the information about the instantaneous period T and cyclic ratio $C = t/t'$ of the input signal based on the pulses detected in the current and voltage paths discussed above. The instantaneous period and cyclic ratio are calculated by a digital counter running on a reference clock and triggered by

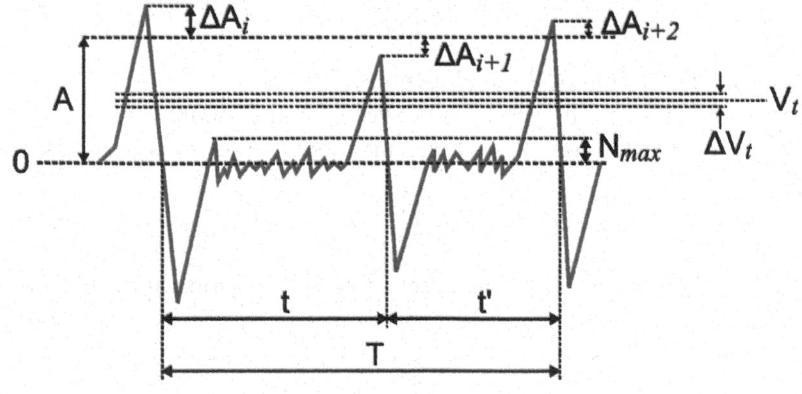

Figure 4.42 Input signal model.

the detected pulses. The reference frequency value is determined based on the cyclic ratio C measurement accuracy requirements. Since C is calculated digitally, t and t' must be measured precisely enough to fulfill the requirements for ΔC.

4.9.2.2 Theoretical performance

To achieve the accuracy of ΔC = 0.0001, $1/\Delta C$ = 10000 periods of the reference signal must fit into the shortest period of the input signal, as shown in the equation:

$$F_{ref} \geq F_{max} \cdot \frac{1}{\Delta C} = 4000\ Hz \cdot 10000 = 40\ MHz \tag{4.2}$$

Given the reference frequency, the accuracy of the frequency measurement is:

$$\Delta F = \frac{F_{max}}{F_{ref}} \cdot 1\ Hz = 0.0001\ Hz, \tag{4.3}$$

where as implied by 1 Hz constant, the precondition is that both the count of F_{ref} and the count of input pulses are integrated for at least 1 s. The same frequency and cyclic ratio accuracy can be achieved with lower clock frequency by averaging the counts over several signal periods. Among the advantages of lower clock frequency is more reliable operation of digital counter and arithmetic unit over wide temperature ranges, larger choise of clock generators (for example, quartz oscillators) etc. We have chosen as specification a value of F_{ref} = 10 MHz for the clock frequency. The minimum counter resolution is defined by F_{min} and F_{ref}, as follows:

$$n_c \geq \log_2(\frac{F_{ref}}{F_{min}}) = \log_2(\frac{40\ MHz}{50\ Hz}) = 19.6096 bits \tag{4.4}$$

A counter with a resolution of 20 bit was implemented.

The threshold voltage should lie in between the maximum noise voltage and minimum positive signal peak level in order to minimize the risk of miscounts:

$$V_t = \frac{(A - max\{\Delta A_i\}) - N_{max}}{2} + N_{max} \tag{4.5}$$

Hysteresis opening ΔV is 10% of the voltage range free of noise and amplitude variations.

$$\Delta V_t = \frac{A(1 - \frac{\Delta A\%}{100\%}) - A\frac{N\%}{100\%}}{10} \tag{4.6}$$

We define the instantaneous frequency error as follows:

$$\Delta F_i = F_{sig,i} - \frac{F_{ref}}{N_i} \tag{4.7}$$

We define the average frequency error as follows:

$$\Delta F_{avg,i} = \frac{\sum_{n=1}^{i} \Delta F_n}{i} \tag{4.8}$$

System simulations were performed with deterministic signal edges as follows:

1. Set initial delay between the signal and clock edges (range: from 0 to $1/F_{ref}$)
2. For each i-th signal cycle determine and plot ΔF_i and $\Delta F_{avg,i}$
3. Repeat steps 1 and 2 for different initial phase

The averaging over 4 cycles is performed to achieve the required accuracy. The average frequency error for a deterministic input signal reduces at every iteration. The speed of error fading depends on signal and reference frequency ratios. When the input signal frequency is an integer multiple of the reference clock frequency, the error remains constant. In real application the sensor signal as well as the reference clock will always contain some jitter resulting in non-monotonic average error fading. After some averaging cycles the error will not decrease any further. For the highest input signal frequency of 4 kHz, the system simulation proves that specified sub-1 Hz frequency error can be achieved after 15 averaging cycles iterations. The error fading can also be observed when the signal value is not an integer multiple of the reference clock frequency. The system averaged frequency error simulation results for deterministic input signal edges are presented in Figure 4.43.

The system averaged frequency error simulation results when jitter with uniform distribution between ± 25 ns was added to the input signal edges are presented in Figure 4.44.

4.9.3 Circuit Design and Implementation

Present section will describe the details of the frequncy signal conditioning circuit design. The top level circuit schematic is presented in Figure 4.45.

4.9.3.1 Input circuit

The sensor signal is applied to the ASIC via off-chip series resistors R_1 and R_2 located in the module, as shown in Figure 4.46. The half-wave rectification

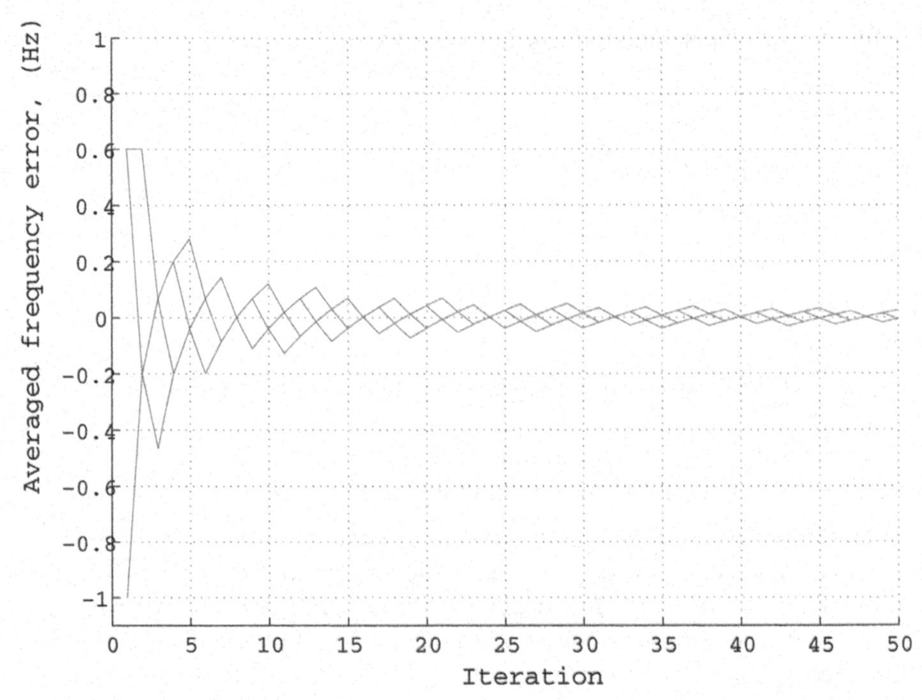

Figure 4.43 System averaged frequency error simulation results for a maximum input frequency F_{sig} = 3999 Hz and a reference clock frequency F_{ref} = 10 MHz.

and pulse detection is performed afterwards. The differential sensor signal is applied to two input circuits: one performing the signal reception and transformation for further processing while the other one being a dummy load for the sensor signal. Two 2.5 kΩ resistors R_5 and R_6 at the input provide differential 5 kΩ impedance required for proper sensor loading. Given the large input voltage amplitudes, resistors R_5 and R_6 are chosen to be high-power surface-mounted device (SMD) components specified for high temperature operation. The *pn* junction diodes, D_1 and D_2, clamp the negative voltage, while positive voltage is converted to current and voltage by means of MOS diodes M_1, M_2 and M_3. Such a configuration of the input circuitry allows for the handling of low-amplitude signals (below 2 V or 4 V_{pp}) using the voltage processing chain, and large-amplitude signals (from 2 V up to 50 V) using the current conversion and detection circuit. The simulated MOS diode current and voltage outputs for multiple temperatures are presented in Figure 4.47. The border between the voltage-to-voltage and voltage-to-current processing lies between 1.5 V and 2 V, for different temperatures. Because of

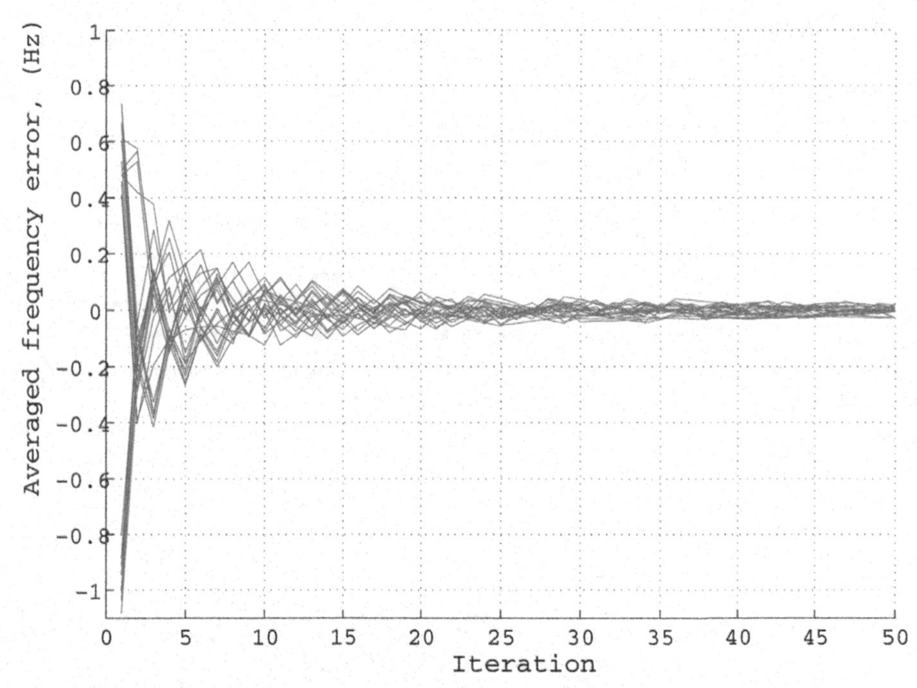

Figure 4.44 System averaged frequency error simulation results with added uniformly distributed jitter between ±25 ns for a maximum input frequency F_{sig} = 3999 Hz and a reference clock frequency F_{ref} = 10 MHz.

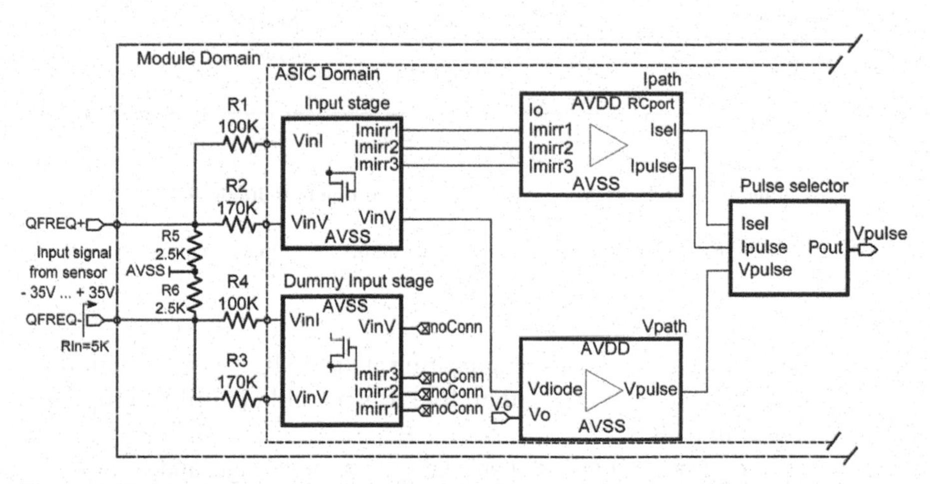

Figure 4.45 Top level circuit schematic of the signal conditioning unit processing the high voltage frequency signal.

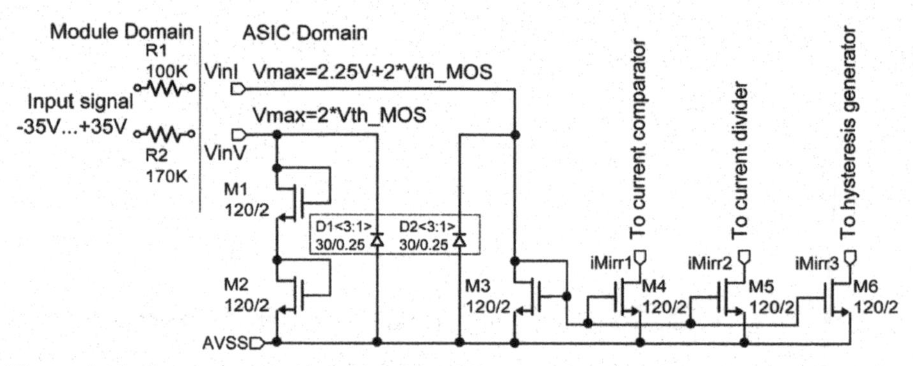

Figure 4.46 Input stage circuit schematic. Diodes D1, D2 are providing the current path for negative input voltage.

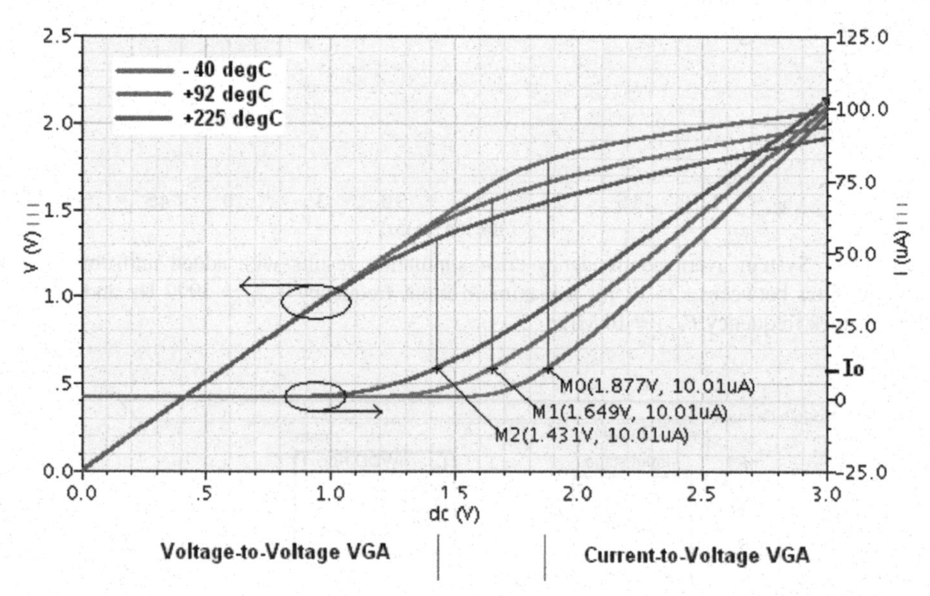

Figure 4.47 Simulated MOS diode current and voltage outputs over input signal for multiple temperatures.

the resistor-diode network, the voltage appearing at the ASIC is always within the supply voltage of the chip (5 V), therefore it is possible to utilize proper ESD protection at all pads.

The pulse detection principle based on peak current or voltage, and variable threshold is presented in Figure 4.48. The minimum peak value is set to prevent false triggering caused by noise. The peak value of the pulses is detected at

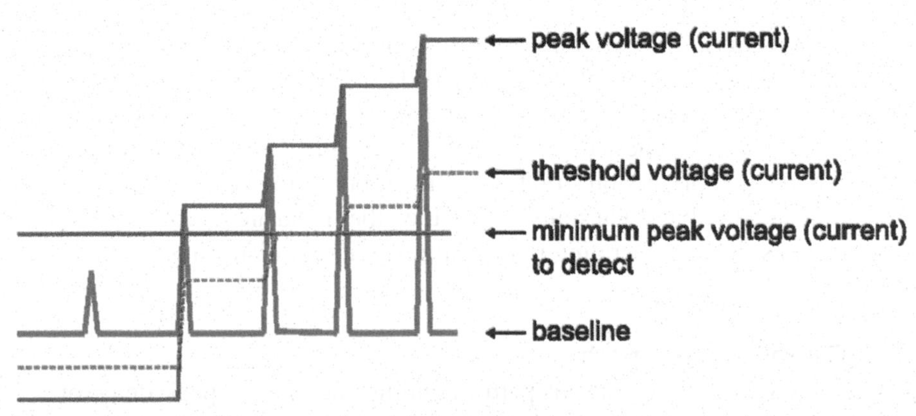

Figure 4.48 Pulse detection principle based on peak current (voltage) and variable threshold.

each cycle. The threshold value is generated from the peak value by dividing it by 2. Once a pulse overshoots the threshold, a digital output pulse is generated. For better noise immunity, the overshoot event is detected by a Schmitt trigger with the hysteresis window positioned around the threshold. The window opening is approximately 10% of the voltage range free of noise and amplitude variations.

The robustness of the signal conditioning unit against ESD event is considered for two handling cases: an electrostatic discharge directly at the ASIC inputs (event occurring during module assembly) and a discharge at the module input. High-ohmic off-chip resistors in series to ASIC inputs and 2.5 kΩ shunt load resistors protect the front-end circuitry from high currents, thus making ESD protection of the complete module very efficient. The Human body model (HBM) includes a 150 pF capacitance charged to ±2 kV and a series 1.5 kΩ resistance [15]. When an ESD event according to HBM occurs at the module input (either differentially or between one input and ground), the R_5 and R_6 high power resistors divide the ±2 kV voltage. The divided voltage is further applied to the ASIC inputs via resistors R_1 and R_2. The MOS and *pn* diodes of the input circuit clamp the voltage at chip inputs keeping it within ±5 V. Because of R_1 and R_2 the ESD current flowing into the ASIC does not exceed 20 mA, which complies with current limits specified for the X-FAB XI10 devices. Similarly, the ESD stress according to machine model (MM), 200 V charged capacitance discharging via a 0.5 μH inductance) does not cause any damage to the chip due to high-ohmic series resistors R_1 and R_2 and a diode-based input circuitry. The ASIC is less ESD immune against the event directly at chip inputs. The input MOS and *pn* diodes provide the first-order protection. The possible overstress or damage may occur due to an excessive voltage or current during an ESD event. According to circuit simulations, the ASIC may directly withstand a stress up to 500 V HBM or up to 100 V MM. The robustness of ASIC inputs against ESD may be improved by placing dedicated protection structures at the inputs which on one hand are able to clamp large voltages/currents and on the other hand are generating low leakage currents at high operating temperatures, without distorting the input signal.

4.9.3.2 Current detect path

The circuit diagram of the current path including the current peak detector is presented in Figure 4.49. The input current is applied to the peak detector formed by the current comparator M_8-R_1, off-chip hold capacitor C_{ext} and the charge pump M_{12}-$OP1$. The current comparator generates the voltage V_{cmp} proportional to the difference between the applied input current and saturation current of M_8. The sensitivity of the current comparator depends on the output impedance of M_8 and the impedance of the input current source r_s. A high sensitivity implies a high value of $r_{ds8}//r_s$. When the input current is higher than the peak current, V_{cmp} will be lower than V_{gs}, and M_{12} will pump current into C_{ext} to equalize the currents. When the input current is lower than the peak current, V_{cmp} will be higher than V_{gs} and the voltage at C_{ext} will not change. Because of this, the charge current for the hold capacitor C_{ext} is decoupled from the input current. This provides large time constants and a fast charge time while having a relatively low input current. The peak current of M_{12} can be several orders of magnitude larger than the input current. $OP1$ is the driver for M_{12}. In this way, the current peak detector is capable of detecting current pulses with sharp edges and low duty cycles. Threshold currents generated by M_9 and M_{10} form the hysteresis for the Schmitt trigger $TS1$, which detects the overshoot of the threshold by the input current pulse. Transistors M_5, M_6, and M_7 set the minimum peak current to be detected. Once the input current pulses exceed the minimum peak current value the output I_{sel} goes high indicating that the current sensing path is active. More details about I_{sel} are provided in Section 4.9.3.4. $OP1$ is a standard operational transconductance amplifier (OTA) [16] with a 2–4 V input common mode voltage range and a gain of 33 dB. The current sensing path was designed for a maximum value of the mirror currents $I_{Mirr1}, I_{Mirr2}, I_{Mirr3}$ of 1 mA, 500 μA and 550 μA, respectively. Resistors R_1–R_4 are reducing the sensitivity to noise voltage at the external RC port, they are limiting the bandwidth of the feedback

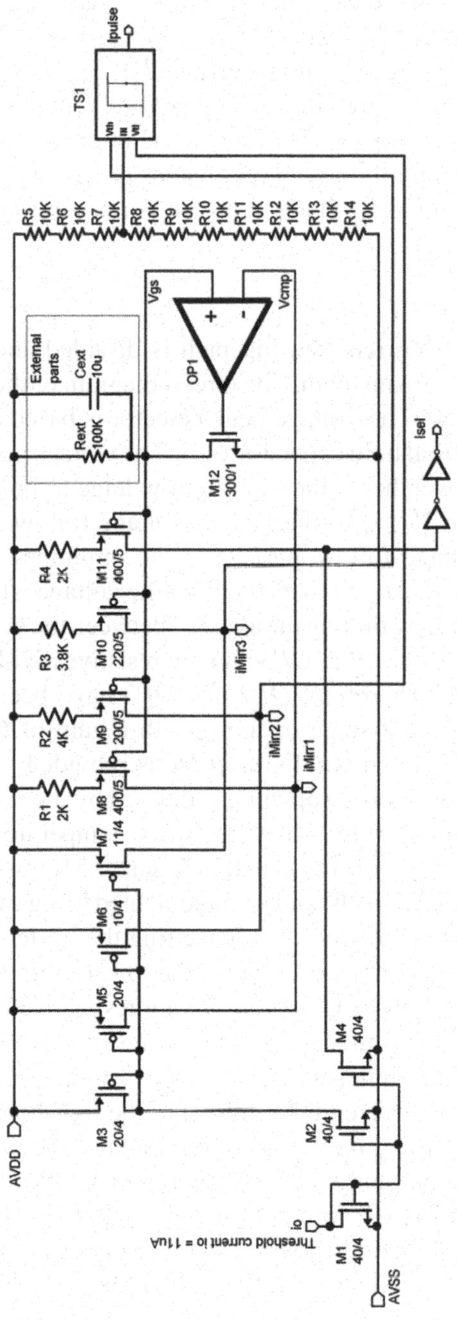

Figure 4.49 Circuit schematic and device sizes of the current sensing path.

circuit and they are increasing the voltage drop at the external R_{ext}, C_{ext} port for large DC currents. Their minimum value is defined by R = 2V/I_{max} = 2 kΩ. The threshold current *io,* the mirror currents I_{Mirr1}, I_{Mirr2}, I_{Mirr3} and the bias currents of *OP*1 and *TS*1 in Figure 4.49 are generated by the bias block presented in [1]. The impact over pulse generation due to temperature and process variation of *io,* I_{Mirr1}, I_{Mirr2} and I_{Mirr3} is minimized by the high gain of the negative feedback network formed by *OP*1, M_{12} and (R_{ext}, C_{ext}).

4.9.3.3 Voltage path

At low input voltages the current sensing path is disabled and detection is performed by the voltage sensing path. The circuit diagram of the voltage path is presented in Figure 4.50. The voltage peak detector is based on the charge pump *OP*1-M_{12} and external hold capacitor C_{ext}. The voltage peak is detected only if the input signal overshoots the minimum voltage to be detected. The input signal is applied at V_{diode} to the PMOS voltage follower M_9-M_8. At the input of another voltage follower M_7-M_6, the minimum voltage to be detected of 0.2 V is applied. M_5-M_4 sets the baseline voltage which is used to generate the threshold level by finding the average between the baseline V_{bl} and peak V_{peak} voltages. OpAmps *OP*2, *OP*3 and the resistive divider $R_1 \ldots R_{10}$ implement this operation. Therefore, *OP*2 and *OP*3 have been designed as NMOS input class-AB output stage opamps, with a gain of 60 dB and an offset voltage of 10 mV. The output digital pulse is provided by the Schmitt trigger *TS*1. The hysteresis window opening is defined by the voltage drop on R_{10} and remains constant over the –40°C to 225°C temperature range. The schematic of the Schmitt trigger is presented in Figure 4.51. It is being used in both voltage and current detect paths and is implemented using two differential amplifiers and a RS-flip-flop. V_{tl} and V_{th} are setting the hysteresis levels for the device. When the input voltage overshoots the lower or upper thresholds, the RS-flip-flop is set to High or Low state by the respective amplifier. The advantage of the used circuit is independent of hysteresis window opening on temperature and temperature-independent current consumption defined solely by the biasing currents for differential amplifiers. The threshold voltages V_{tl} and V_{th} are dynamically changing with the amplitude of the input signal, as implemented in both current and voltage sensing blocks. While the voltage drop of D_1–D_3 *pn*-diodes is 0.3 to 0.4 V smaller than the voltage drop on a MOS connected diode, the diode voltage drop dependence on temperature matches the minimum input voltage requirements of the *OP*1 differential pair

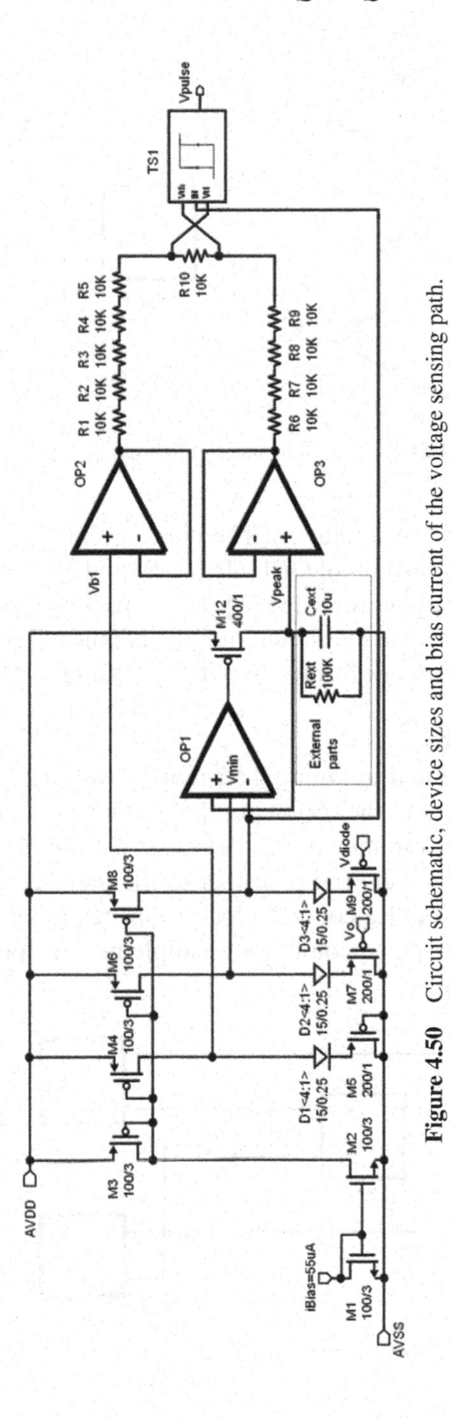

Figure 4.50 Circuit schematic, device sizes and bias current of the voltage sensing path.

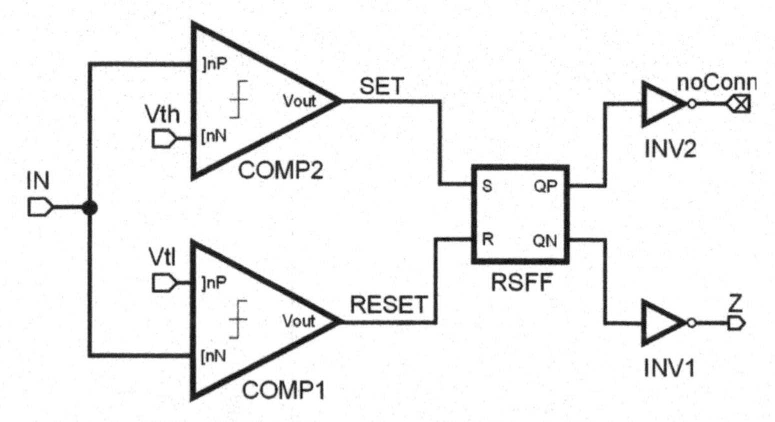

Figure 4.51 Circuit schematic of Schmitt Trigger.

to keep its current source in saturation. The bias current i_{Bias} of the voltage sensing path and the bias currents of $OP1$, $OP2$, $OP3$ and $TS1$ in Figure 4.50 are generated by the bias block presented in [1]. The impact over pulse generation due to temperature and process variation of i_{Bias} is minimized by the gain of the negative feedback network formed by $OP1$, M_{12} and (R_{ext}, C_{ext}).

4.9.3.4 Pulse selector

The pulse selector presented in Figure 4.52 represents the interface between the voltage/current path and the following digital pulse counter. It validates either the output signal from the voltage detection path in Figure 4.49 or the current detection path 13 and send it further to the pulse counter. Based on the incoming pulses, the pulse counter block measures the frequency of the input sensor signal. The pulse counter was implemented together with the

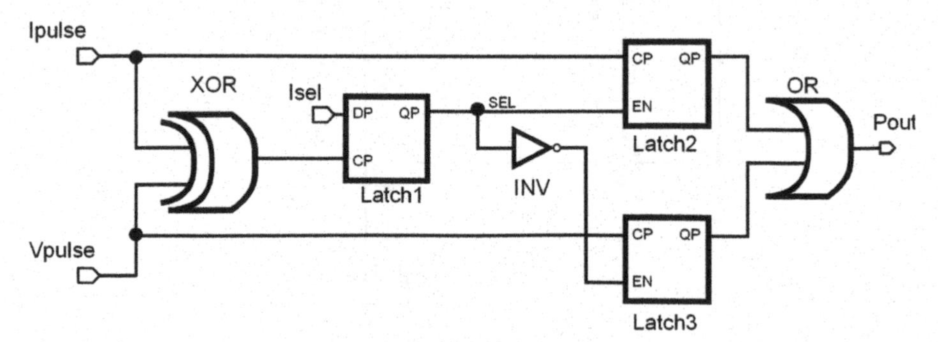

Figure 4.52 Pulse selector schematic.

ARINC429 transmitter in the digital flow, and is beyond the scope of this section.

The I_{sel} input indicates which one of the signal paths will be chosen. Two latched gated clock cells *Latch*2 and *Latch*3, [17] are bypassing either I_{pulse} or V_{pulse} depending on the selection state. The change of selection state in-between the rising edges of I_{pulse} or V_{pulse}, namely t_1 and t_2, is presented in Figure 4.53. If selection signal I_{sel} goes from low to high during t_1 neither I_{pulse} or V_{pulse} can generate the pulse at P_{out}. A no-triggering-error at the output P_{out} when both signal paths are active ($I_{pulse} = V_{pulse}$) is prevented by *XOR* and *Latch*1 latching the I_{sel} control signal. In [3], the simulation results of an applied 4 kHz input signal and the detected pulses at the output with a width of 14 μs are presented.

4.9.4 Experimental Results

As mentioned in Section 4.9.2, the design principle of the HIGHTECS module was based on a custom SOI ASIC being used for the signal conditioning and signal processing from the range of sensors (i.e. temperature, strain gauges, frequency), multiplexing, analogue to digital conversion and transmission of data through an ARINC 429 databus. The HIGHTECS module helps reducing the weight of the helicopter engine by at least several kilograms [18]. The HIGHTECS ASIC was then integrated with voltage regulators, resistors and capacitors onto a ceramic hybrid circuit. The ceramic hybrid circuit was assembled in a hermetic Kovar package and hermetically sealed in an inert gas atmosphere. The Kovar package was then mounted into a stainless steel enclosure, as shown in Figure 4.54. In addition to the hybrid circuit, a high temperature printed circuit board (PCB) containing the high power SMD

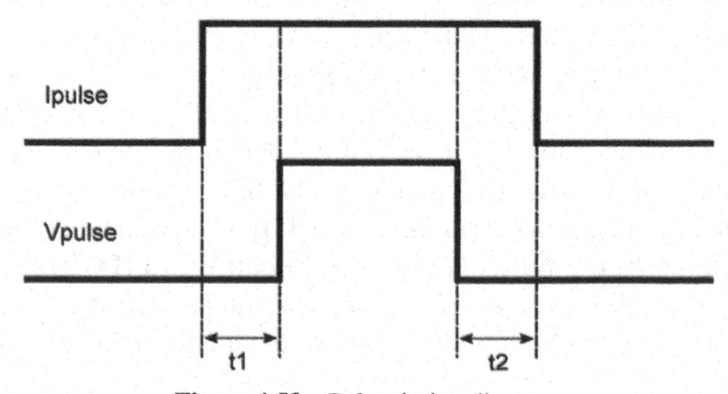

Figure 4.53 Pulse timing diagram.

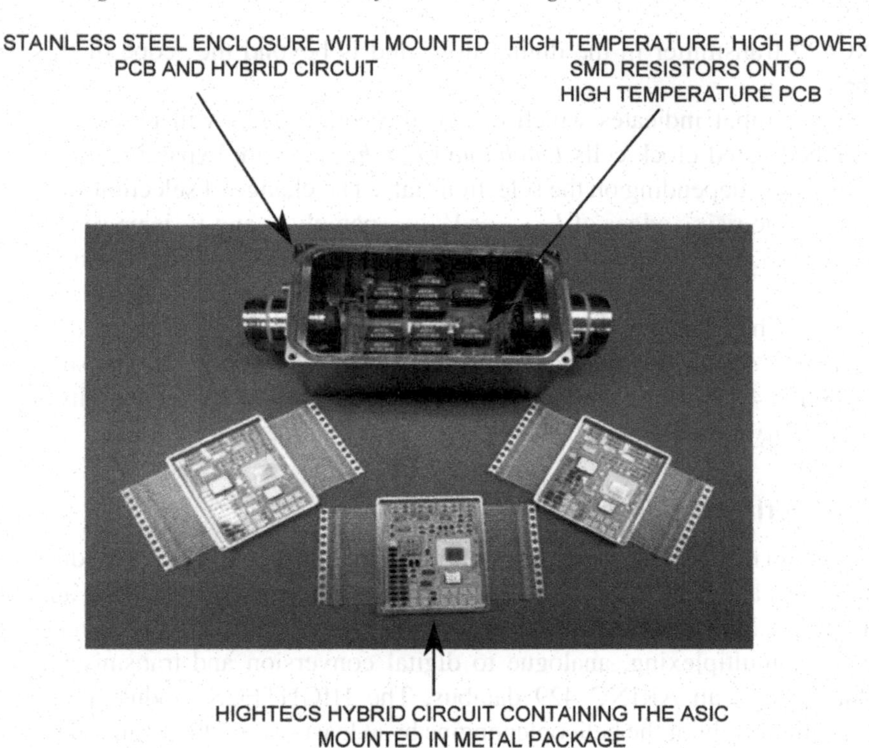

Figure 4.54 Stainless steel enclosure with mounted PCB and hybrid circuit including the HIGHTECS ASIC.

resistors for the frequency signal conditioning unit in Section 4.9.3.1 was designed and manufactured. This board was also mounted in the HIGHTECS module. The connections between the leads on the metal package, connection pads on the PCB and the connectors in the stainless steel enclosure were made with polymide insulated copper wire. Several measurement results of the HIGHTECS hybrid have been published in [19].

In order to minimize the high temperature adverse effects such as material decomposition or phase change, increased stress due to coefficient of thermal expansion mismatches and leakage, the HIGHTECS ASIC was assembled in a 181-pin high temperature co-fired ceramic (HTCC) package and mounted onto a custom high temperature polyimide printed circuit board (PCB) to enable testing up to 235°C. The measurement results of the packaged HIGHTECS ASIC are reported below. The high temperature evaluation PCB is presented in Figure 4.55.

The proposed signal conditioning unit was fabricated in the 1 μm X-FAB SOI CMOS process. Figure 4.56 shows the layout of the frequency signal

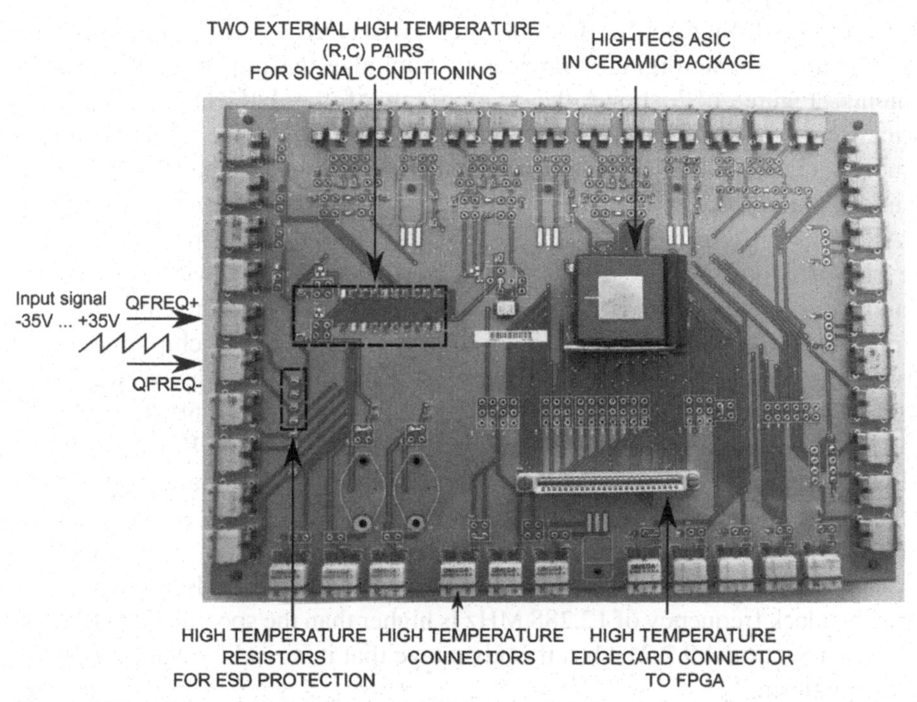

Figure 4.55 Photograph of the customized high temperature evaluation board used during HIGHTECS ASIC characterization measurements.

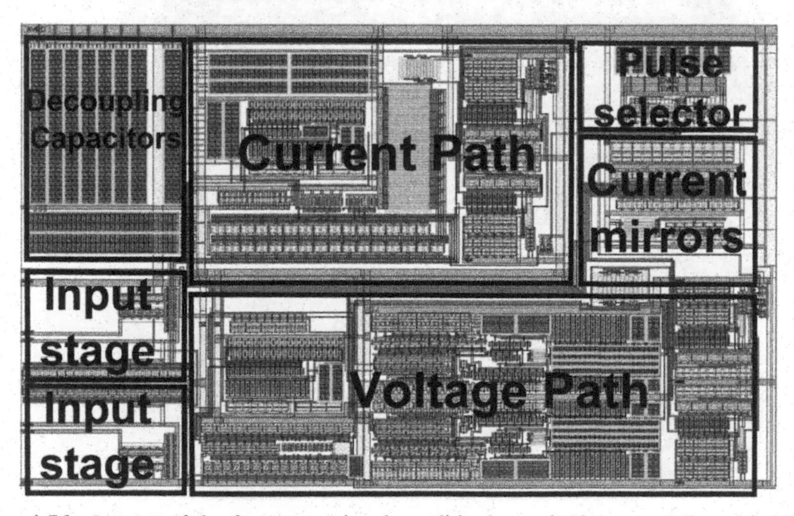

Figure 4.56 Layout of the frequency signal conditioning unit [1 *mm* × 1.5 *mm*] integrated onto the fabricated HIGHTECS ASIC.

conditioning unit included in the fabricated HIGHTECS ASIC. All devices were layouted as interdigitated devices to minimize the impact of thermal gradients. Figure 4.57 shows the micrograph of the HIGHTECS ASIC fabricated in the X-FAB XI10 SOI process [1]. The active area of the frequency signal conditioning unit, excluding the bonding pads and I/O drivers, is approximately 1.5 mm^2. The system level test platform to measure the ARINC words from HIGHTECS ASIC are shown in Figure 4.58. The core circuit is powered by a 5 V analog and digital supply with separate grounds. For further noise reduction, decoupling high temperature capacitors are placed between power supply and grounds in the unused chip area. A software procedure to read the ARINC word via ChipScope was written in VHDL. This corresponds to the JTAG to USB bus in Figure 4.58. A digital receiver able to read the digital ARINC 429 output of the ASIC has been designed and implemented on a Xilinx XC3S500E FPGA (Spartan3E family) placed on a GODIL programming board. The GODIL board is providing a 49.152 MHz reference signal which is divided internally by 4 to generate a clock frequency of 12.288 MHz for the digital processor pulse counters. While the clock frequency of 12.288 MHz is higher than the specified value of 10 MHz in Section 4.9.2.2, it has the advantage that it is readily available on the test platform.

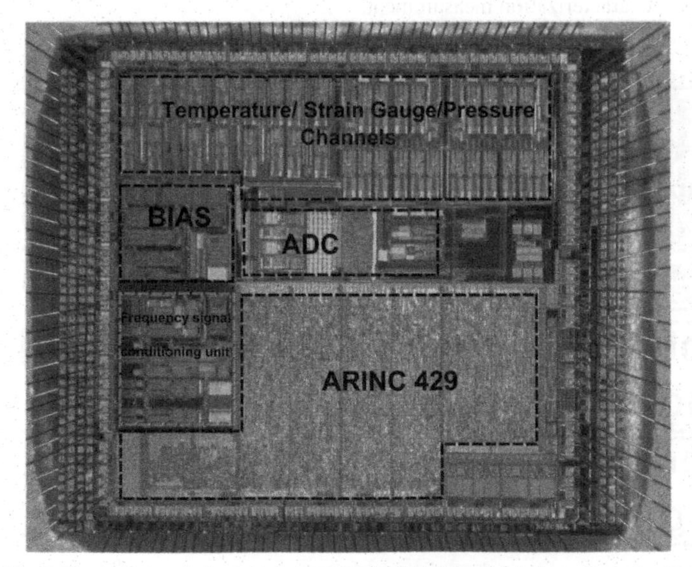

Figure 4.57 Micrograph of the bonded HIGHTECS ASIC fabricated in the X-FAB XI10 SOI process. The frequency signal conditioning unit is positioned on the left side the ASIC (in blue).

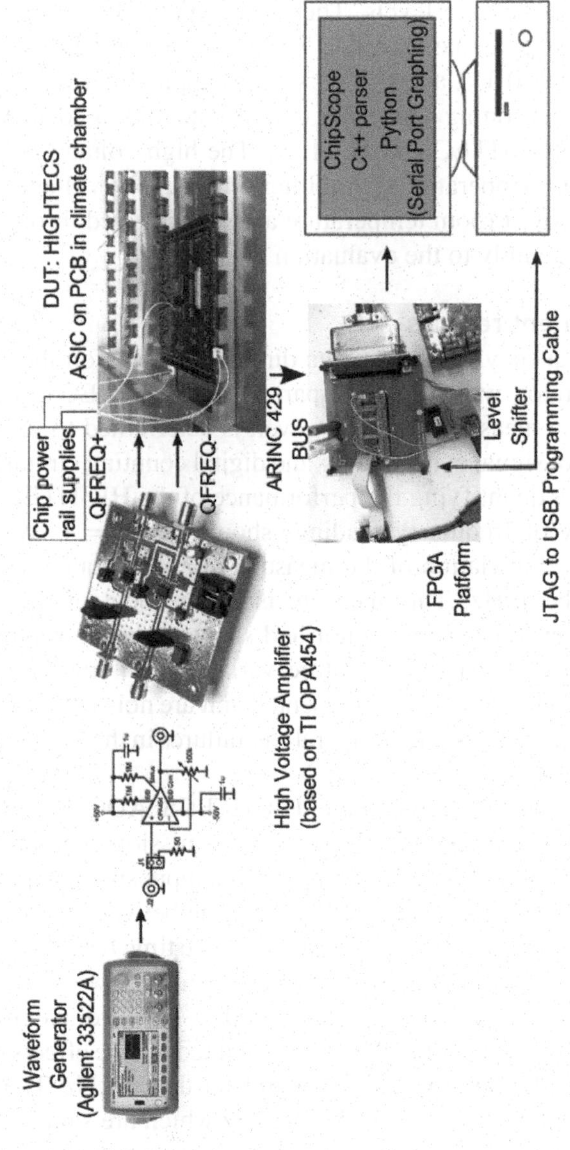

Figure 4.58 Block diagram of the HIGHTECS ASIC hardware & software test platform.

4.9.4.1 High voltage amplifier

The input signal was generated using a dual-channel waveform generator followed by a high voltage amplifier. An Agilent 33522A waveform generators provides a ±5 V peak voltage levels. Therefore, an additional high voltage amplifier is required to generate the V_{in} = 70 V_{pp} input signal. During testing, a sawtooth wave with 70 V_{pp} and 0 V offset was applied at the input of the frequency signal conditioning unit. The output data was read in from the ARINC data bus of HIGHTECS ASIC [1, 3]. The high voltage amplifier is based on the TI OP454 operational amplifier with supply voltage of up to ±50 V. It operates only at room temperature and is connected through a high temperature cable assembly to the evaluation board.

4.9.4.2 Measurement results

The accuracy of frequency measurements directly depends on the accuracy of a reference generator, which is not a part of the HIGHTECS ASIC. The linearity of frequency measurements, however, is determined to most extent by the performance of signal conditioning and digital counting circuitry and is used as a measure for quantifying the performance of the HIGHTECS ASIC. Ideally, the linearity of frequency readings should be R^2 = 1. We specify the maximum allowed deviation of the measured output frequency from the linear trendline to be 1 Hz. This measure is independent of the absolute frequency of a reference clock generator and shows only the robustness of the HIGHTECS ASIC against temperature variation. Among the potential causes for linearity degradation in the presented unit are noise in the front-end circuitry, thresholds deviation over temperature, failures in the digital counters caused by temperature change etc.

The digital processor following the HIGHTECS ADC is performing the pulse counting for the ARINC 429 serial transmission. The processor's counters measure two periods t and t' of a sawtooth input signal which are the times between the sensor signal pulses: t is the time between loaded phonic wheel pulse and unloaded phonic wheel pulse; t' is the time between unloaded phonic wheel and loaded phonic wheel pulse. There are two sensor signal pulses for each cycle of the engine. The periods $t + t'$ are averaged over 4 cycles to improve the measurement accuracy and reduce the effects of noise. An error flag is set if the input frequency is greater than 4 kHz. The t and t' timers are 20 bit counters clocked at 12.288 MHz which are saturating when the maximum count is reached. The digital pulses coming out of the frequency signal conditioning unit are obtained by reading out the ARINC 429 serial bus interface. The frequency of the input signal was swept between 50 Hz to 4 kHz across the 25°C–235°C temperature range to measure if the pulse readings and linearity varied across temperature. Measurement results are indicating the linearity of the output pulse counting measurements was very high (>0.01%) and did not vary over temperature. Measurement results at 25°C and 235°C are presented in Figures 4.59 and 4.60, respectively. The measurements were limited at 235°C due to the maximum temperature specification of the PCB mounted capacitors and of the polyimide material of the PCB.

There are virtually no differences between the measurement results at 25°C and those at the maximum temperature of 235°C, which shows the temperature stability of the design. Based on the measured R^2 linearity values, a maximum averaged frequency error of 0.3 Hz was calculated which is fullfilling the frequency error requirement Table 4.10 presents the specification versus measurement results.

The linearity values of the output pulses counting versus temperature are presented in Figure 4.61. Measurement results of several HIGHTECS ASICs are indicating same linearity values for the first past comma seven digits.

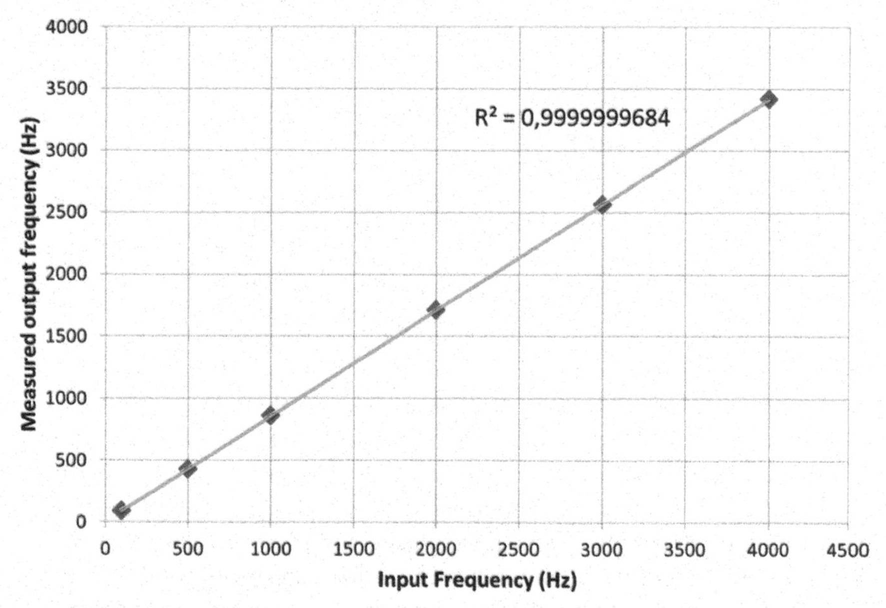

Figure 4.59 Measured output frequency (red dots) via ARINC and FPGA at 25°C shows a linearity value of $R^2 = 0.9999999684$ with a reference clock frequency of $F_{ref} = 12.288$ MHz.

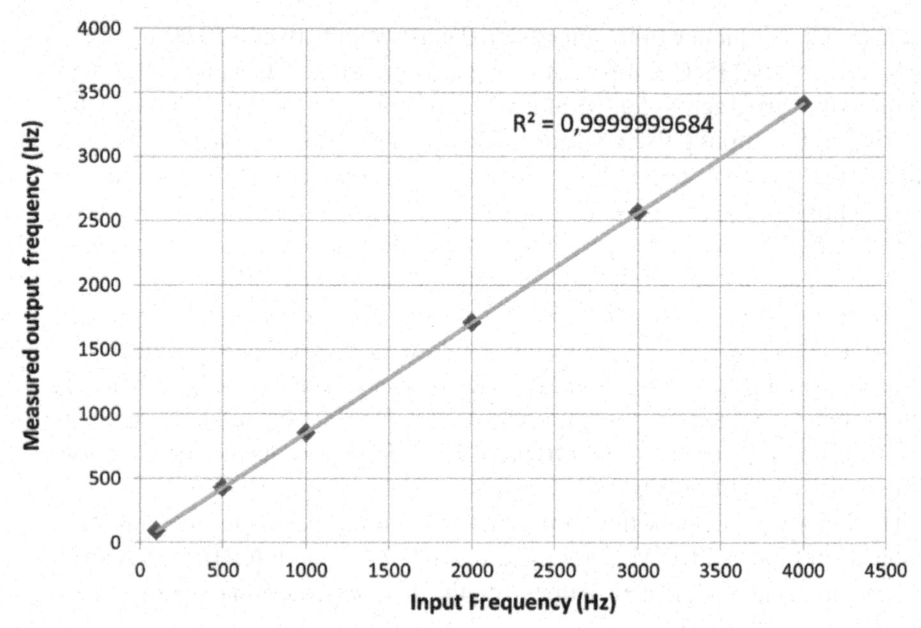

Figure 4.60 Measured output frequency (red dots) via ARINC and FPGA at 235°C shows a linearity value of $R^2 = 0.9999999684$ with a reference clock frequency of $F_{ref} = 12.288$ MHz.

Table 4.10 Specification versus measurement results

Parameter	Specification	Measurement	Units
Temperature	200	235	$^{\circ}C$
F_{min}	50	50	Hz
F_{max}	4000	4000	Hz
V_{in}	70	70	V
$\Delta F_{avg,i}$	<1	<0.3	Hz
F_{ref}	10	12.288	MHz

The presented signal conditioning unit is a discrete-domain conditioning system which detects the frequency of a periodical input signal. The frequency detection is free of errors as the front-end circuitry is correctly detecting the rising/falling edges of the input signal using internally-generated threshold voltages. Even though the absolute values of threshold voltages/currents vary slightly with temperature, the temperature variations are low enough for the error-free detection of input signal edges, and thus frequency. The time jitter caused by temperature-dependent thresholds is well below the period of the $F_{ref} = 12.288$ MHz reference clock.

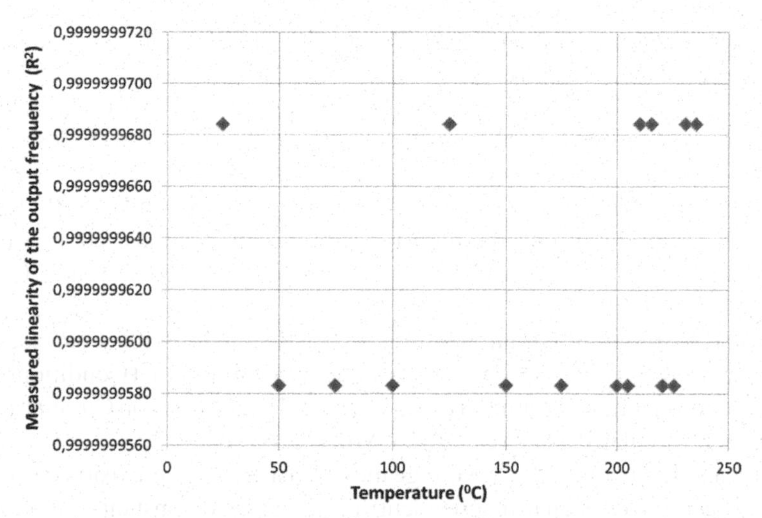

Figure 4.61 Measured linearity values (R^2) of the output frequency over the 25°C to 235°C temperature range are within specification limits. The reference clock frequency value is F_{ref} = 12.288 MHz.

References

[1] L. Stoica, V. Solomko, T. Baumheinrich, R. D. Regno, R. Beigh, S. Riches, I. White, G. Rickard, and P. Williams, "Design of a high temperature signal conditioning ASIC for engine control systems – HIGHTECS," in *Proc. IEEE International Symposium on Circuits and Systems*, Melbourne, Australia, May. 2014, pp. 2117–2120.

[2] R. Hogervorst, "Design of a low-voltage low-power CMOS operational amplifier cells," Ph.D. dissertation, Univ. of Delft, Delft, The Netherlands, Jun. 1996.

[3] L. Stoica, V. Solomko, T. Baumheinrich, R. D. Regno, R. Beigh, I. White, G. Rickard, and P. Williams, "Design of a frequency signal conditioning unit applied to rotating systems in high temperature aero engine control," in *Proc. IEEE International Symposium on Circuits and Systems*, Lisbon, Portugal, May. 2015, pp. 1154–1157.

[4] D. Flandre, L. Demeus, V. Dessard, A. Viviani, B. Gentinne, and J. P. Eggermont, "Design and application of SOI CMOS OTAs for hightemperature environments," in *Proc. IEEE 24th European Solid-State Circuit Conference*, The Hague, The Netherlands, Sep. 1998, pp. 404–407.

[5] R. Hogervorst, J. Tero, R. Eschauzier, and J. Huijsing, "A compact power-efficient 3V CMOS rail-to-rail input/output operational amplifier for VLSI cell libraries," *IEEE J. Solid-State Circuits*, vol. 29, pp. 1505–1513, Dec. 1994.

[6] J. Huijsing, *Operational amplifiers.* Dordrecht, The Netherlands: Springer Science+Business Media B.V., 2011.

[7] Y. Lam, and W. Ki, "CMOS bandgap references with self-biased symmetrically matched current-voltage mirror and extension of Sub-1-V design," IEEE Trans. VLSI Sys., vol. 18, no. 6, pp. 857–865, June 2010.

[8] L. Stoica; R. Ghandi; Cheng-Po Chen; E. Andarawis; V. Solomko; S. Riches, A signal conditioning unit for high temperature applications, 2016 IEEE International Symposium on Circuits and Systems (ISCAS), Year: 2016, Pages: 2403–2406.

[9] L. Stoica; V. Solomko; T. Baumheinrich; R. Del Regno; R. Beigh; G. Rickard; P. Williams; S. Riches, "A High Temperature Frequency Signal Conditioning Unit for Aeronautical Rotating Systems", IEEE Transactions on Circuits and Systems I: Regular Papers Year: 2016, Volume: 63, Issue: 5.

[10] K. Koli and K. Halonen, "Low voltage mos-transistor-only precision current peak detector with signal independent discharge time constant," in *Proc. IEEE International Symposium on Circuits and Systems*, Hong Kong, PRC, Jun. 1997, pp. 1992–1995.

[11] C. Mead, R. Lyon, and R. Sarpeshkar, "A low-power wide dynmic range analog VLSI cochlea," *Analog Integrated Circuits and Signal Processing*, vol. 16, pp. 245–274, Aug. 1998.

[12] R. Sehgal, A. Singh, and W. Serdijn, "CMOS ultra low-power wavelet filter based sense amplifier for cardiac signal analysis," in *Proc. 19th Annual Workshop on Circuits, Systems and Signal Processing (ProRisc)*, Veldhoven, The Netherlands, Nov. 2008, pp. 260–266.

[13] S. Zhak, M. Baker, and R. Sarpeshkar, "A low-power wide dynamic range envelope detector," *IEEE J. Solid-State Circuits*, vol. 38, pp. 1750–1753, Oct. 2003.

[14] ARINC429. (2015, Sep.) ARINC 429 Overview. [Online]. Available: https://www.aim-online.com/pdf/OVIEW429.PDF

[15] M. Ker, J. Peng, and H. Jiang, "ESD test methods on integrated circuits: an overview," in *Proc. IEEE International Conference on Electronics Circuits and Systems*, Valetta, Malta, Sep. 2001, pp. 1011–1014.

[16] R. Baker, *CMOS Circuit Design, Layout and Simulation.* Hoboken, New Jersey: John Wiley & Sons Inc., 2008.

[17] J. Yuan and C. Svensson, "High-speed CMOS circuit technique," *IEEE J. Solid-State Circuits*, vol. 24, pp. 62–70, Feb. 1989.

[18] J. Rigaud and L. Stoica, "Turbomeca TCON," Oct. 2015, Meeting Notes.

[19] S. Riches, C. Warn, K. Cannon, G. Rickard, and L. Stoica, "Design and assembly of high temperature distributed aero-engine control system demonstrator," in *Proc. International Conference on High Temperature Electronics (HiTEC)*, Albuquerque, United States, May. 2014.

5

Characterization of Prototypes

5.1 Assessment of Prototype Performance

The assessment of the HIGHTECS prototype parts has covered the following components:

- ASIC in PGA Package
- Hybrid Circuit
- High Temperature PCB for resistors
- HIGHTECS Module

An initial assessment of prototype SiC Transient Voltage Suppressors (TVS) devices in lightning tests has also been carried out.

Long term tests have also been performed on a SOI test chip to identify degradation mechanisms that may occur during the lifetime of the product.

5.1.1 ASIC in PGA Package

90 HIGHTECS ASICs were manufactured in 181 I/O PGA packages to enable functional and environmental tests to be carried out. A high temperature printed circuit board has been designed and manufactured to enable characterisation tests to be carried out.

5.1.2 Functional Tests

Functional testing of the HIGHTECS ASIC assembled in the PGA package has been carried out for the following blocks.

- Bias network
- Single ended to differential converter
- T1 channel measurements
- Band gap voltage
- Strain gauge bridge channels; SG11, SG12, P3

- T4 channel measurement
- Tfo1 and Tfo2
- ADC

The results have shown that the performance of the functional blocks was broadly in line with the expected performance from simulation. The HIGHTECS ASIC as designed has been also shown to function through to the generation of the dual output ARINC 429 data.

The dual outputs from the ARINC 429 databus on the HIGHTECS ASIC were connected to an AIM UK APU 429-4 2 channel transmitter/2 channel receiver to ARINC 429 interface, see Figure 5.1. The data transmitted was then handled by an AIM UK PBA.pro-ARINC429 Database Manager Component. Representative output data are shown in Figure 5.2.

There are several areas of ASIC performance that need further attention, including the ADC, Tfo2 signal, and Nfreq.

5.1.3 Design Changes Implemented for 2nd Version of HIGHTECS ASIC

The work on modifications to the ADC design was undertaken in two stages; the first through modification to the top metallisation layers on 1st version of the HIGHTECS ASIC and the second through a complete re-design of the mask set to incorporate the changes to the ADC and to Tfo2 and Nfreq VHDL code.

Figure 5.1 HIGHTECS ASIC in PGA package connected to ARINC 429 data reader.

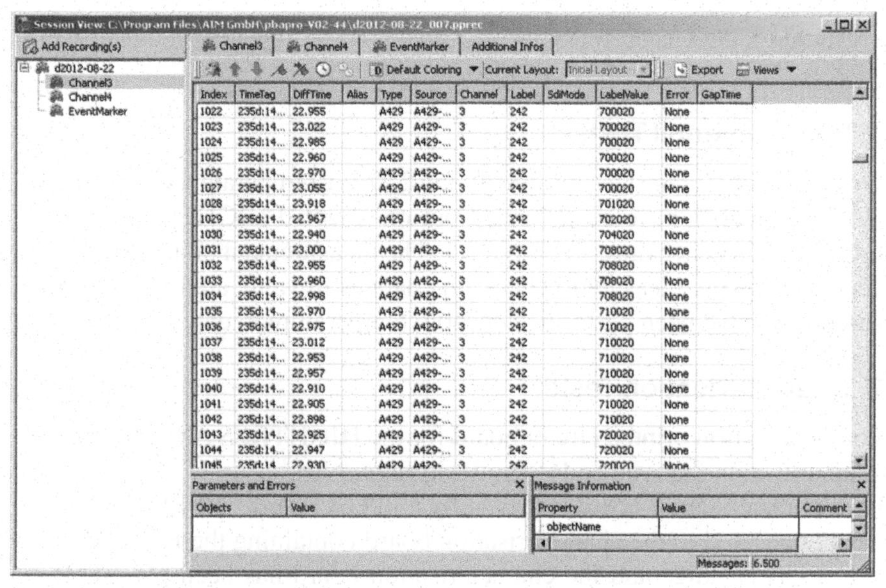

Index	TimeTag	DiffTime	Alias	Type	Source	Channel	Label	SdiMode	LabelValue	Error	GapTime
1022	235d:14...	22.955		A429	A429-...	3	242		700020	None	
1023	235d:14...	23.022		A429	A429-...	3	242		700020	None	
1024	235d:14...	22.985		A429	A429-...	3	242		700020	None	
1025	235d:14...	22.960		A429	A429-...	3	242		700020	None	
1026	235d:14...	22.970		A429	A429-...	3	242		700020	None	
1027	235d:14...	23.055		A429	A429-...	3	242		700020	None	
1028	235d:14...	23.918		A429	A429-...	3	242		701020	None	
1029	235d:14...	22.967		A429	A429-...	3	242		702020	None	
1030	235d:14...	22.940		A429	A429-...	3	242		704020	None	
1031	235d:14...	23.000		A429	A429-...	3	242		708020	None	
1032	235d:14...	22.955		A429	A429-...	3	242		708020	None	
1033	235d:14...	22.960		A429	A429-...	3	242		708020	None	
1034	235d:14...	22.998		A429	A429-...	3	242		708020	None	
1035	235d:14...	22.970		A429	A429-...	3	242		710020	None	
1036	235d:14...	22.975		A429	A429-...	3	242		710020	None	
1037	235d:14...	23.012		A429	A429-...	3	242		710020	None	
1038	235d:14...	22.953		A429	A429-...	3	242		710020	None	
1039	235d:14...	22.957		A429	A429-...	3	242		710020	None	
1040	235d:14...	22.910		A429	A429-...	3	242		710020	None	
1041	235d:14...	22.905		A429	A429-...	3	242		710020	None	
1042	235d:14...	22.898		A429	A429-...	3	242		710020	None	
1043	235d:14...	22.925		A429	A429-...	3	242		720020	None	
1044	235d:14...	22.947		A429	A429-...	3	242		720020	None	
1045	235d:14	22.930		A429	A429-	3	242		720020	None	

Figure 5.2 ARINC 429 output from HIGHTECS ASIC.

5.1.4 Modification to Top Layer Metallisations on 1st Version of HIGHTECS ASIC

The following changes were implemented in the re-layout of the top layer metallisations on 3 off wafers of the 1st version of the HIGHTECS ASIC held at pre-poly stage:

- Create separate Vrefp and Vrefn inputs of the ADC
- Re-route metallisation lines that were crossing ADC
- Connect DP and DQ to digital pads
- Vdd and Gnd connections of the comparators decoupled from those of the other digital part by direct routing to the pads

The wafers were manufactured at X-FAB and were then wafer probed using the ADC functionality test. The results showed that non-linearity of the ADC output was still apparent at Vdda 5 V and 5.5 V, but was linear at 6 V.

5.1.5 2nd Version of HIGHTECS ASIC

Based on the above results, the 2nd version of the HIGHTECS ASIC was re-designed and manufactured. The ASIC wafer was probe tested and selected samples which passed the within the limits set on the linearity of ADC functionality were assembled into PGA packages. The results of testing the

ASICs in PGA packages at Vdda: 5.5 V for ADC linearity is shown in Figure 5.3.

5.1.6 Environmental Tests

High temperature storage tests (at 200°C and 250°C), temperature cycling and shock/vibration tests have been carried out on selected HIGHTECS ASICs assembled in PGA packages, see Table 5.1. The measurement of analogue I_{dd} has been used as the measure to check on changes in value after testing. All measurements have been performed at room temperature to date.

5.1.7 Characterisation Tests

The characterisation printed circuit board for the HIGHTECS ASIC has been designed and manufactured and is shown in Figure 5.4.

The characterisation board is driven by a FPGA board. The FPGA board has been connected to the characterisation board containing the HIGHTECS ASIC in PGA package and the characterisation board has been placed into either an oven for high temperature testing up to 275°C or a chamber for testing down to –40°C. Characterisation tests have been carried out on the 1st and 2nd version of the ASIC; a couple of example results are presented below.

The change in voltage bandgap against temperature up to 250°C and the effective temperature coefficient are shown in Figure 5.5. SG2 output from the HIGHTECS module at +225°C is shown in Figure 5.6.

5.1.8 Prototype SiC Transient Voltage Suppressors

Prototype SiC Transient Voltage Suppressor (TVS) devices were assembled with copper tags providing the conductor path, as shown in Figure 5.7. These SiC devices can operate at temperatures of at least 200°C, which is above the temperature of commercial Si based lightning protection devices. Preliminary lightning testing has been carried out on these devices, following the procedures of DO-160E.

Pin injection lightning tests following the procedures detailed in RTCA (Radio Technical Commission for Aeronautics)/DO-160E, Section 22.5.1 – Lightning Induced Transient Susceptibility were carried out on the waveforms and levels shown in Table 5.2.

The prototype devices were shown to clamp successfully at the levels and waveforms highlighted in Table 5.2. Further work to assess leakage currents and incorporate the devices into circuits is required.

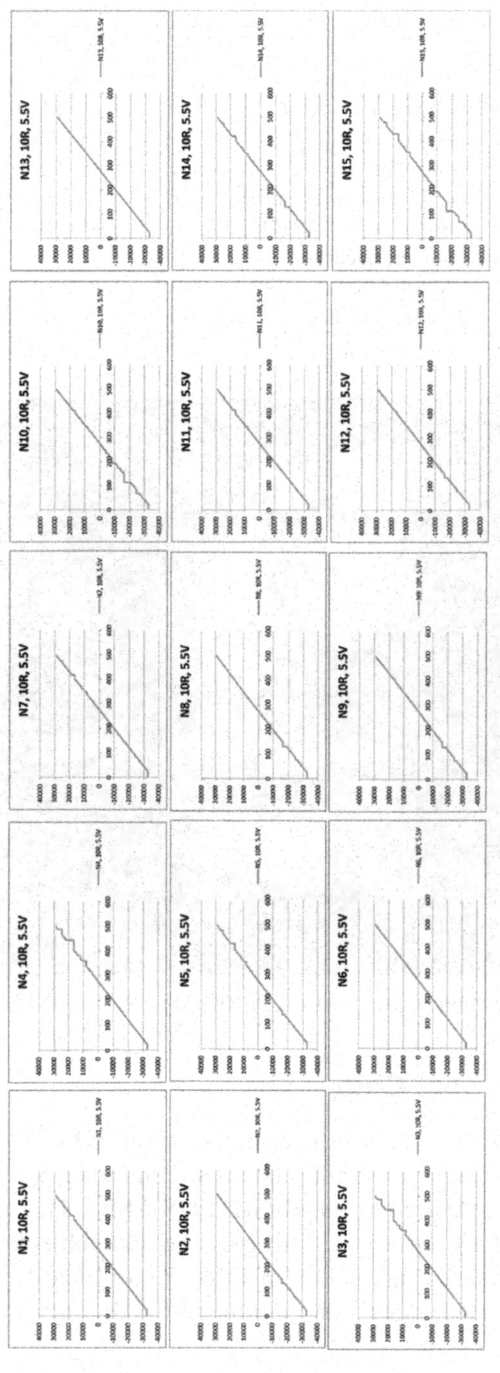

Figure 5.3 ADC linearity plot of 2nd version of HIGHTECS ASIC assembled in PGA packages.

Table 5.1 Summary of environmental tests on HIGHTECS ASIC in PGA package

Environmental Test	Test Condition	Average % Change in I_{dd} Current
High Temperature Storage	8000 hours at 200°C	–2.48%
High Temperature Storage	8000 hours at 250°C	–5.27%
Temperature Cycling	–40°C to +250°C, 375 cycles (min 3 hours at each limit)	–5.39%
Vibration	Random 10–2000 Hz, 0.1 g/Hz^2 3 hours each axis	+1.4%
Vibration and Shock	As vibration test above + Shock 1500 g, 0.5 ms, 5 times, 5 axes	+1.2%

Figure 5.4 Characterisation board for testing of HIGHTECS ASIC in PGA package.

5.2 Testing of SOI Test Chip

Environmental tests have been performed throughout the HIGHTECS project on a SOI test chip fabricated using the same semiconductor process (X-FAB XI10 1 μm) in the manufacture of the HIGHTECS ASIC. This test chip has been assembled into a 48 pin HTCC DIL package and subjected to various environmental tests as described below in Table 5.3 to identify degradation mechanisms that may occur during the lifetime of the product.

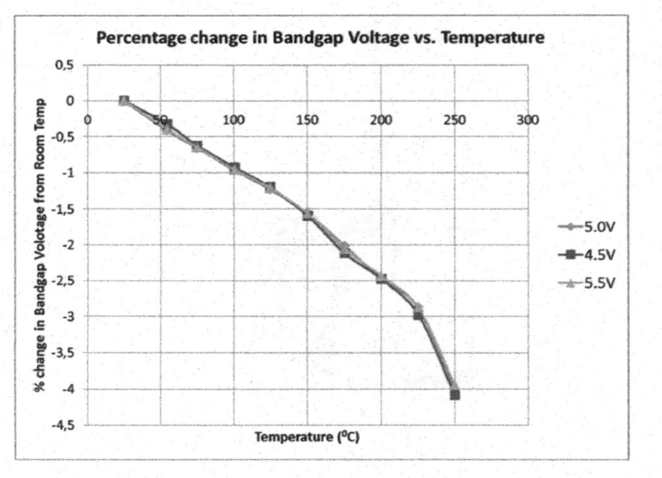

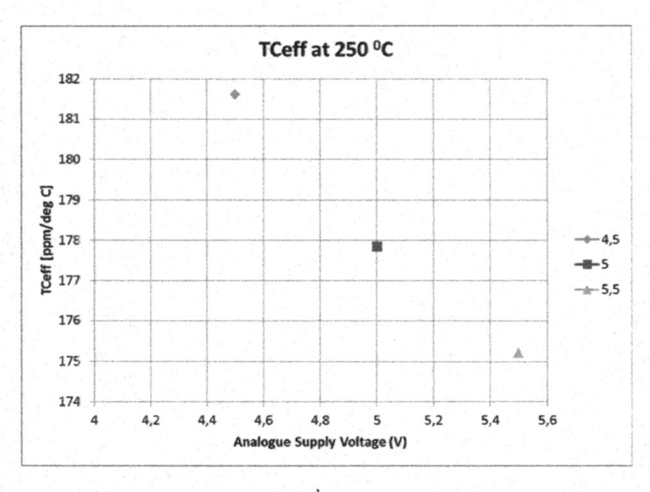

Figure 5.5 Voltage bandgap change with temperature and effective temperature coefficient of 2nd version of HIGHTECS ASIC.

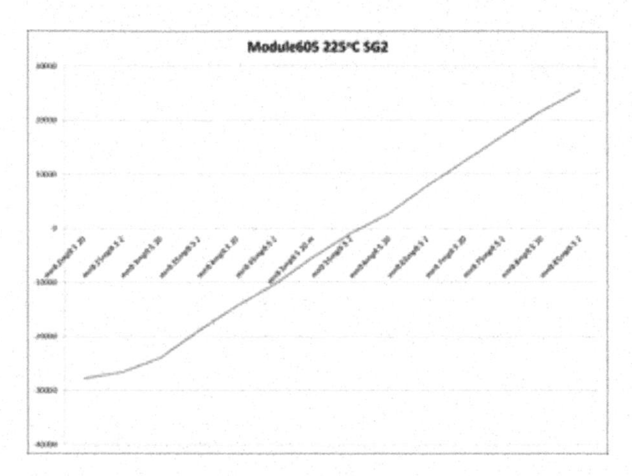

Figure 5.6 Output from SG2 sensor on HIGHTECS module at +225°C.

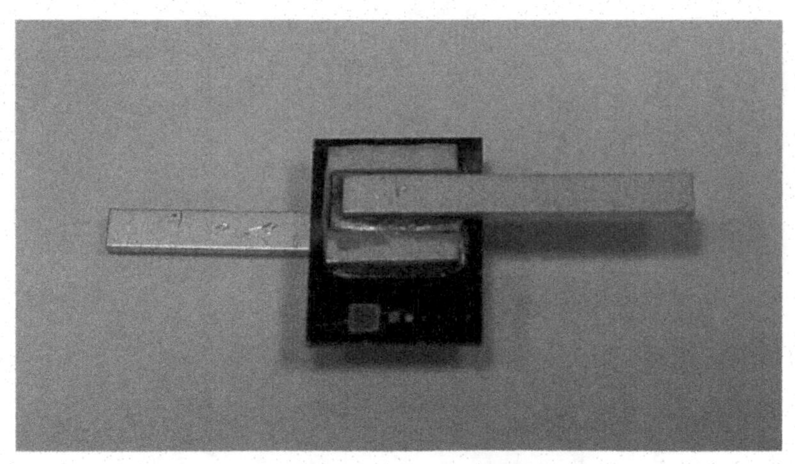

Figure 5.7 Prototype SiC TVS devices with copper tags attached.

5.2.1 High Temperature Storage (250°C)

The main limiting factor on the performance of the SOI test devices assembled in HTCC packages when exposed to temperatures of 250°C for up to 11,000

Table 5.2 Lightning induced transient susceptibility – pin injection tests

	Waveforms		
	3	4	5A
Level	Voc/Isc	Voc/Isc	Voc/Isc
1	100/4	50/10	50/50
2	250/10	125/25	125/125
3	600/24	300/60	300/300
4	1500/60	750/150	750/750
5	3200/128	1600/320	1600/1600

Notes: Voc – peak open circuit voltage (V), Isc – peak short circuit current (A).
For Waveform 3, the frequency was 1 MHz.

Table 5.3 Summary of environmental tests carried out on SOI test chip

Environmental Test	Test Duration	SOI Test Chip Details
High Temperature Storage (250°C)	11088 hours	Batch 1 – Au/TiW metallisation, bond out options 1–6
	7056 hours	Batch 2 – Au/Pd/Ni metallisation, bond out options 1–6
Rapid Thermal Cycling (–40°C to +225°C)	2680 cycles @ ~5 mins per cycle	Batch 3 – Au/Pd/Ni metallisation, bond out option 7
Vibration (Room Temperature and 200°C)	Resonance Random Sine	Batch 3 – Au/Pd/Ni metallisation, bond out option 7

hours appeared to be the packaging materials used in the assembly process rather than the device itself. There were two principal sources for the degradation; firstly, through the formation of intermetallics between the Au wire bond and the Al metallisation on the bond pad, despite the presence of over bond pad metallisations designed to prevent diffusion of the Al metallisation, and secondly, through the deterioration of the high temperature die attach adhesive within the hermetically sealed package, which caused the formation of whiskers around the bond pads, see Figure 5.8.

In the case of wire bonding, Al-1%Si wire wedge bonded to the Al metallisation on the device was selected as the most stable option. In the case of the die attach, an inorganic Au-Si eutectic solder is recommended to avoid problems of deterioration of organic materials.

5.2.2 Rapid Temperature Cycling (–40°C to +225°C)

The following temperature profiles were provided by Turbomeca:

5.2.2.1 Profile No. 1: 4 × following cycle

- 30 mins, power on, on ground, with stopped engine (Temperature –50°C to +50°C)
- 90 min, power on, in flight with running engine (Temperature 150°C)
- 30 mins, power on, on ground, with stopped engine (Temperature 250°C – approx. 4 mins at 250°C, with 26 mins cooling to ambient)
- 210 mins, power off, on ground, with stopped engine

5.2.2.2 Profile No. 2: 2 × following cycle

- 613 mins, power off, on ground, with stopped engine (Temperature –50°C to +50°C)
- 4 mins, power on, on ground, with stopped engine (Temperature –50°C to +50°C)
- 99 mins, power on, in flight with running engine (Temperature 150°C)
- 4 mins, power on, on ground, with stopped engine (Temperature 250°C)

The number of cycles were calculated for each profile against the specified target operating lifetime of 50,000 hrs. This equates to 8,333 cycles under Profile No. 1 and 4,167 cycles under Profile No. 2. To accelerate this number of cycles in reduced time it was proposed to ramp from the minimum and maximum temperature extremes at the maximum cooling/heating rate with

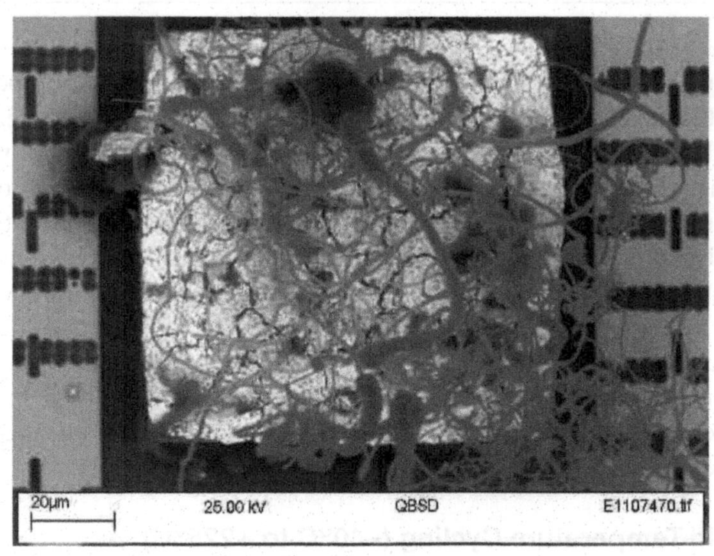

Figure 5.8 SEM picture of unbonded bond pad of adhesive bonded SOI device after 11,088 hours exposure to 250°C showing growth of whiskers.

no dwell time at any temperature. Examples of a full day equivalent running are shown in Figures 5.9 and 5.10 for both profiles.

By having no high temperature dwell any failures due to accumulated strain (e.g. in solder joints and die attach) will be accelerated even more since there will be no opportunity afforded for annealing at temperature.

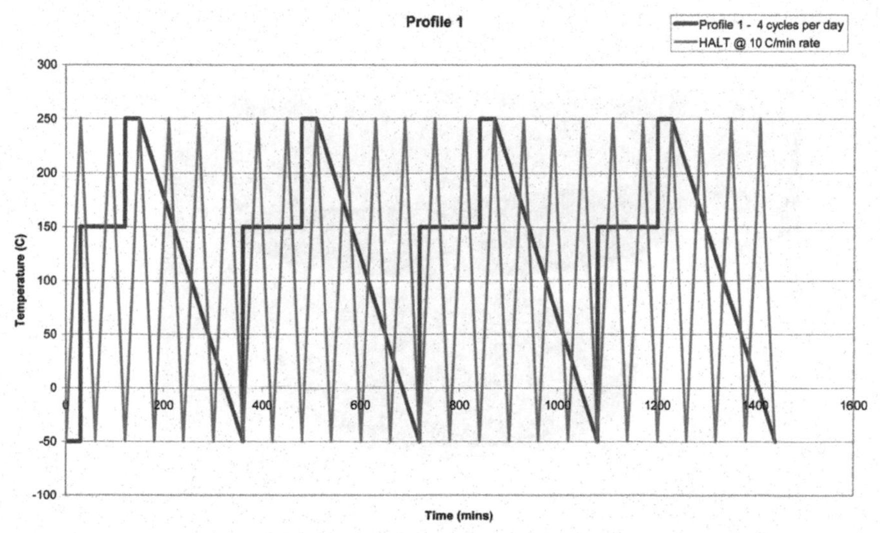

Figure 5.9 Example of full day equivalent running for Profile 1.

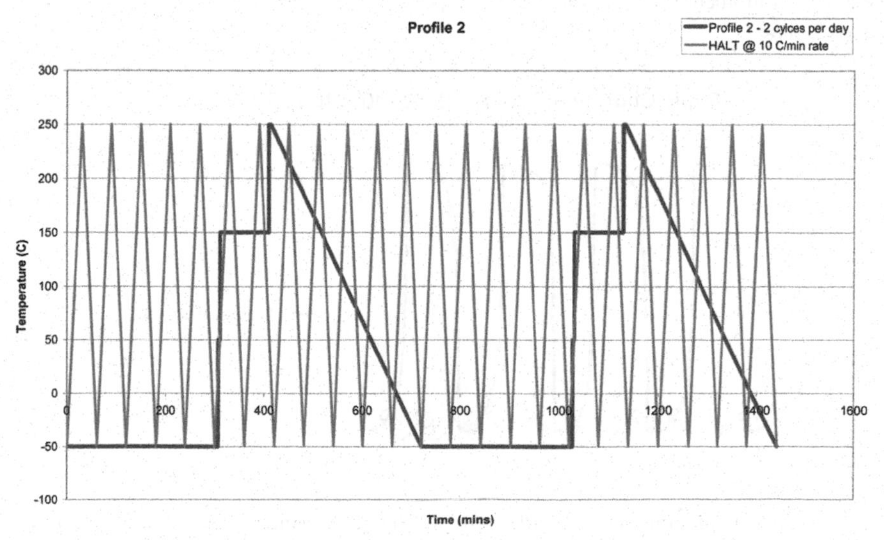

Figure 5.10 Example of full day equivalent running for Profile 2.

Trials have been carried out to assess the performance of SOI test chips after exposure to rapid change of temperature from –40°C to +225°C over a periodic cycle time of ~5 minutes which was the minimum cycle time that could be achieved with the equipment available, see Figure 5.11 for the thermal cycling equipment and Figure 5.12 for the temperature profile.

Packaged SOI test chips were run using rapid thermal cycling and monitoring the gain of an op-amp contained within the device before and after testing.

Figure 5.11 Equipment for rapid change of temperature from –40°C to +225°C with 320 second cycle time.

Rapid Change of Temperature -40°C to 225°C

250
200
150
100
50
0
-50
-100

0 120 240 360 480 600 720 840 960 1080 1200 1320

Temperature, oC

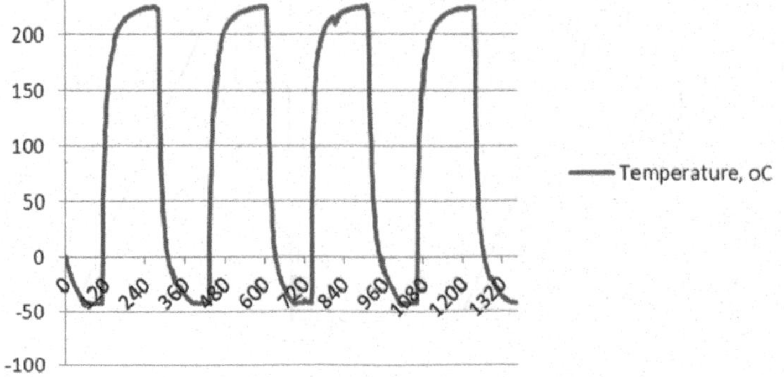

Figure 5.12 Measured temperature profile for rapid change of temperature with 320 second cycle time.

A set of 3 off samples was submitted to 2680 cycles, which represents nearly a third of the expected number of thermal cycles during the lifetime of the component for temperature profile 1 and then tested again. The results showed that little obvious change in op-amp gain was observed. Visual inspection of this sample did not show any obvious degradation.

5.2.3 Vibration (Room Temperature and 200°C)

Vibration testing has been undertaken to ensure that the die attach and wire bonds are not affected by any resonance effects in the sinusoidal vibration modes and random vibrations. The results of the electrical tests after vibration testing showed little difference indicating that conventional vibration testing should not be major concern to the reliability of the assembled ASIC. However vibration testing at temperature may cause additional issues and some additional tests have been carried out to assess this aspect.

SOI test devices with op-amp functional blocks were assembled into HTCC packages with 25 μm diameter Au wire. The gain of the op-amps was measured prior to testing and again after random vibration testing at a temperature of 25°C and 200°C. The results showing the change in op-amp gain at test temperatures of 25°C, 125°C and 250°C indicated that there was little discernible difference in the op-amp gain between the vibration tests carried out at 25°C and 200°C.

5.2.4 Silicon Capacitors

Silicon capacitors supplied by Ipdia were analysed using scanning electron microscopy and energy dispersive X-Ray (EDX) analysis before and after thermal ageing at 250°C. The analysis showed that the capacitors were composed of a monolithic piece of silicon with an Al doped guard ring around the active device, with Au flashed Ni plating as the contact metallisation on the connecting pads. After thermal ageing at 250°C for 24 hours in air, the nickel on the contact metallisation has diffused through the Au metallisation and oxidised to a thickness of ~10 nm of NiO.

10 nF, 100 nF and 1 μF silicon capacitors have been incorporated into the hybrid circuit design. Thermal cycling from –40°C to +250°C of 1206 1 μF silicon capacitors onto alumina substrates using a Ag loaded high temperature conductive adhesive have shown some shorting of the capacitors after less than 10 cycles. An initial investigation has been carried out, which has shown some cracking in the contact metallisation, the dielectric and the silicon. It is believed that the stress caused by thermal cycling of the surface mounted capacitors

onto the alumina substrate results in the cracking of the dielectric beneath the contact pads. Alternative options for assembly have been reviewed and trials have been carried out on wire bonded versions of the Ipdia capacitors, which have shown little variation in capacitance, leakage currents and no occurrence of shorts when subjected to temperature storage at 200°C for >3500 hours and 165 cycles from –40°C to +200°C.

5.3 Functional Tests on Eagle Test Systems

5.3.1 Room Temperature Testing

Details of the testing of the HIGHTECS ASIC assembled in the PGA package for the following functional blocks are presented below.

- Bias network
- Single ended to differential converter
- T1 channel measurements
- Band gap voltage
- Strain gauge bridge channels; SG11, SG12, P3
- T4 channel measurement
- Tfo1 and Tfo2
- ADC

The HIGHTECS ASIC as designed has been shown to function through to the generation of the dual output ARINC 429 data, but there are several areas of ASIC performance that need further attention, including the ADC, Tfo2 signal, Nfreq and the repeatability of the band gap voltage measurement.

ADC: The linearity of the ADC output has been shown to depend on the applied voltage and temperature, with some devices performing better than others. It is believed that impedance on the ADC test pad cells may affect the ADC output. Trials have been carried out to isolate the test pad cells from the circuit using a Focused Ion Beam (FIB), but this did not show any difference in the ADC output. An improvement in the linearity of the ADC output was observed when the digital and analogue V_{dd} were separated by 0.5 V, with 5.5 V on V_{dd} and 6.0 V on V_{dda}. Further simulations have been carried out by IMMS to investigate the effect of supply line resistance, which did not show exactly the same behaviour as the measured devices.

From the initial assessment of ADC performance, it was believed there was some variability in the output of different devices, such that some devices may operate at lower voltages than others. A test program to assess the ADC functionality with a maximum tolerance voltage band of ±0.6 V around the

ADC linear output voltage was developed for probe testing at the wafer level. This program identified a small number of devices (7.5% of the wafer) from the wafer that had an ADC transfer function within the tolerance band at an analogue voltage of 5.5 V and digital output of 5 V. These devices will be assembled into the HIGHTECS hybrid circuits to assess their performance. Although these devices have a functioning ADC, the output may not be linear.

It is believed that a modification to the layout of the connections to and the tracking around the ADC is required to reduce the sensitivity to the applied voltage which will require a new mask set and re-spin of the ASIC.

Tfo1/Tfo2 Signal: Due to an error in the VHDL code, the Tfo2 sensor is a repeat of the Tfo1 sensor and will not show in the ARINC 429 message. This can be corrected by changing the VHDL code, which would require a respin of the ASIC.

Nfreq: In the Nfreq module, there is a requirement for a Nfreq pulse before the state machine can change state. This works for frequencies below the minimum, but the state machine becomes stuck of the Nfreq frequency is zero. A change in the VHDL code to overcome this effect is required.

Band Gap Voltage: The repeatability of the band gap voltage measurement appears to be related to the test equipment set up and the application of power resources to the component during testing.

The dual outputs from the ARINC 429 databus on the HIGHTECS ASIC were connected to an AIM UK APU 429-4 2 channel transmitter/2 channel receiver to ARINC 429 interface, see Figure 5.13. The data transmitted

Figure 5.13 HIGHTECS ASIC in PGA package connected to ARINC 429 data reader.

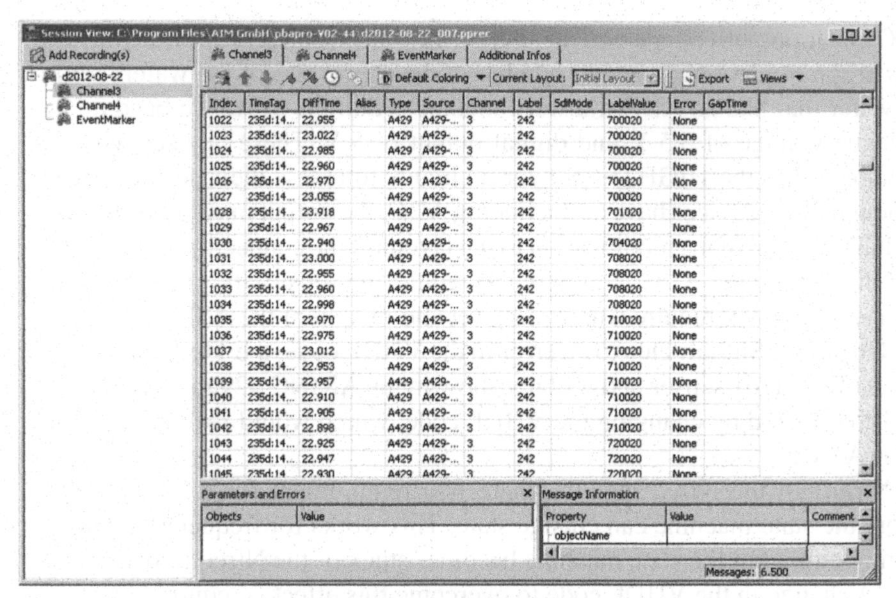

Index	TimeTag	DiffTime	Alias	Type	Source	Channel	Label	SdiMode	LabelValue	Error	GapTime
1022	235d:14...	22.955		A429	A429-...	3	242		700020	None	
1023	235d:14...	23.022		A429	A429-...	3	242		700020	None	
1024	235d:14...	22.985		A429	A429-...	3	242		700020	None	
1025	235d:14...	22.960		A429	A429-...	3	242		700020	None	
1026	235d:14...	22.970		A429	A429-...	3	242		700020	None	
1027	235d:14...	23.055		A429	A429-...	3	242		700020	None	
1028	235d:14...	23.918		A429	A429-...	3	242		701020	None	
1029	235d:14...	22.967		A429	A429-...	3	242		702020	None	
1030	235d:14...	22.940		A429	A429-...	3	242		704020	None	
1031	235d:14...	23.000		A429	A429-...	3	242		708020	None	
1032	235d:14...	22.955		A429	A429-...	3	242		708020	None	
1033	235d:14...	22.960		A429	A429-...	3	242		708020	None	
1034	235d:14...	22.998		A429	A429-...	3	242		708020	None	
1035	235d:14...	22.970		A429	A429-...	3	242		710020	None	
1036	235d:14...	22.975		A429	A429-...	3	242		710020	None	
1037	235d:14...	23.012		A429	A429-...	3	242		710020	None	
1038	235d:14...	22.953		A429	A429-...	3	242		710020	None	
1039	235d:14...	22.957		A429	A429-...	3	242		710020	None	
1040	235d:14...	22.910		A429	A429-...	3	242		710020	None	
1041	235d:14...	22.905		A429	A429-...	3	242		710020	None	
1042	235d:14...	22.898		A429	A429-...	3	242		710020	None	
1043	235d:14...	22.925		A429	A429-...	3	242		720020	None	
1044	235d:14...	22.947		A429	A429-...	3	242		720020	None	
1045	235d:14	22.930		A429	A429-	3	242		720020	None	

Figure 5.14 ARINC 429 output from HIGHTECS ASIC.

was then handled by an AIM UK PBA.pro-ARINC429 Database Manager Component. Representative output data are shown in Figure 5.14.

5.3.2 High and Low Temperature Testing

High and low temperature functional testing of the HIGHTECS ASIC is to be carried out after the completion of the functional tests at room temperature.

5.3.3 Environmental Tests

5.3.3.1 High temperature storage (200°C and 250°C)

High temperature storage tests have been carried out on selected HIGHTECS ASICs assembled in PGA packages. The measurement of analogue I_{dd} has been used as the measure to check on changes in value after testing. All measurements have been performed at room temperature to date. The results for 200°C storage and 250°C storage up to 8000 hours are presented in Tables 5.4 and 5.5 respectively.

Scanning Electron Microscopy (SEM) of thermally aged samples at 200°C and 250°C for 8000 hours has been carried out, which showed little degradation of the HIGHTECS ASIC, the die attach, the wire bond interconnections and the PGA package.

Table 5.4 Temperature storage tests at 200°C on HIGHTECS ASIC in PGA package

		Analogue I_{dd}, mA			
Sample No	Storage Test Temperature	0 Hours	360 Hours	2100 Hours	8000 Hours
74	200°C	10.15	10.35	10.10	10.00
75	200°C	10.26	10.05	9.95	9.91
76	200°C	10.33	10.31	10.33	10.30
77	200°C	10.34	10.29	10.12	9.93
78	200°C	10.25	10.38	10.02	9.91
79	200°C	10.05	10.17	9.99	9.86
80	200°C	10.50	10.44	10.46	10.37
81	200°C	10.25	10.20	10.19	10.10
82	200°C	10.38	10.38	10.20	10.17
83	200°C	10.26	10.37	10.19	10.13
84	200°C	10.35	10.34	9.98	9.95
85	200°C	10.19	10.21	9.89	
86	200°C	10.35	10.34	10.05	10.03
87	200°C	10.27	10.28	9.87	9.77
88	200°C	10.26	10.17	9.77	

Table 5.5 Temperature storage tests at 250°C on HIGHTECS ASIC in PGA package

		Analogue I_{dd}, mA			
Sample No	Storage Test Temperature	0 Hours	260 Hours	2000 Hours	8000 Hours
52	250°C	10.35	10.33	9.83	9.96
53	250°C	10.09	10.12	9.75	9.38
54	250°C	10.02	9.86	9.81	9.58
55	250°C	10.17	10.10	9.92	9.89
56	250°C	10.62	10.39	10.24	9.58
57	250°C	10.33	10.34	9.81	
58	250°C	10.23	10.18	9.74	

5.3.3.2 Temperature cycling (–40°C to +250°C)

Temperature cycling tests have been carried out on selected HIGHTECS ASICs assembled in PGA packages. The measurement of analogue I_{dd} has been used as the measure to check on changes in value after testing. All measurements have been performed at room temperature to date. The results are presented in Table 5.6.

Scanning Electron Microscopy (SEM) of thermally cycled samples from –40°C to 250°C for 375 cycles has been carried out, which showed some cracking of the die attach material, see Figure 5.15.

Table 5.6 Temperature cycling tests from –40°C to 250°C on HIGHTECS ASIC in PGA package

		Analogue I_{dd}, mA			
Sample No	Temperature Cycling Range	0 Cycles	10 Cycles	100 Cycles	375 Cycles
59	–40°C to 250°C	10.07	10.21	9.95	9.54
60	–40°C to 250°C	10.03	10.06	9.74	
61	–40°C to 250°C	9.95	10.09	9.71	
62	–40°C to 250°C	10.39	10.44	10.23	
63	–40°C to 250°C	10.17	10.38	9.96	9.58
64	–40°C to 250°C	10.15	10.15	9.77	9.68
65	–40°C to 250°C	10.22	10.16	9.81	
Average Percentage Change in Analogue I_{dd} Current				–2.56%	–5.39%

Notes on test conditions: Samples stored for at least 3 hours at each temperature extreme within maximum transfer time of 30 minutes.

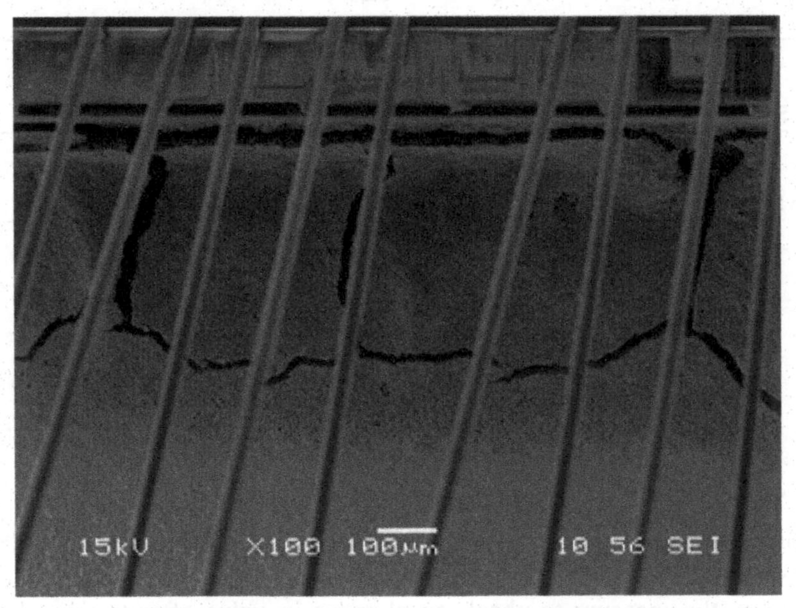

Figure 5.15 Cracking of die attach material after 375 cycles from –40°C to +250°C.

5.3.3.3 Vibration/Shock

Vibration and shock tests have been carried out on selected HIGHTECS ASICs assembled in PGA packages. The measurement of analogue I_{dd} has been used as the measure to check on changes in value after testing. The results are presented in Table 5.7.

Table 5.7 Vibration and shock tests on HIGHTECS ASIC in PGA package

		Analogue I_{dd}, mA		
Sample No	Test	Before	After	% Change
67	Vibration and Shock	10.36	10.52	+1.5
68	Vibration and Shock	10.30	10.51	+2.0
69	Vibration and Shock	10.10	10.27	+1.6
70	Vibration and Shock	10.44	10.42	–0.2
71	Vibration	10.25	10.41	+1.6
72	Vibration	10.21	10.27	+0.6
73	Vibration	10.34	10.55	+2.0

Notes on test conditions:
Vibration test: Random 10–2000 Hz, 0.1 g/Hz^2, 3 hours each axis.
Shock test: 1500 g, 0.5 ms, 5 times, 5 axes.

5.3.3.4 Testing of HIGHTECS hybrid circuit and high temperature PCB containing resistors

Boxes for testing of the HIGHTECS hybrid circuit and HIGHTECS module have been designed and manufactured, as shown in Figures 5.16–5.18. The boxes have been designed to test the various inputs on the HIGHTECS circuit. The hybrid circuit can be mounted onto the socket and tested prior to final assembly.

The output from a HIGHTECS hybrid circuit (Hybrid Serial Number 10511832) with all sensor inputs open circuit connected to the ARINC 429 Bus Monitor User Interface is presented in Figure 5.19. The red LEDs are all indicating open circuit errors as expected. The Tfo2 signal has no error

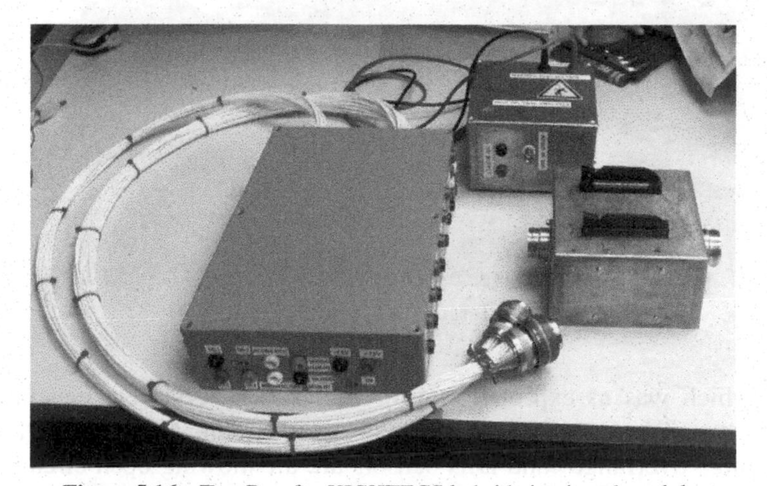

Figure 5.16 Test Box for HIGHTECS hybrid circuit and module.

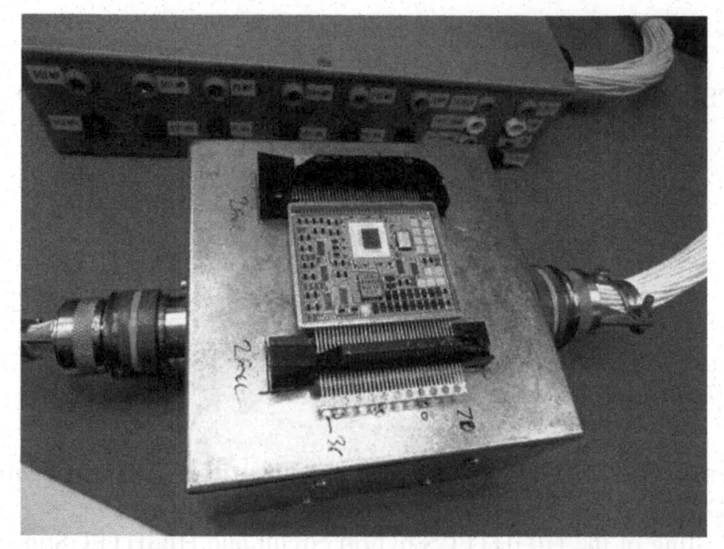

Figure 5.17 Test of HIGHTECS hybrid circuit.

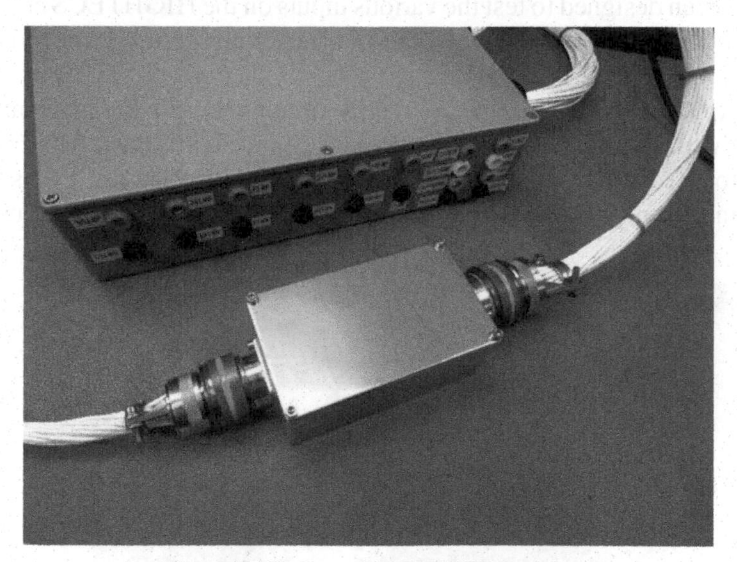

Figure 5.18 Test of HIGHTECS module.

message, which was as expected as no ARINC messages were generated for this sensor. T4 signal is indicating an open circuit. Nfreq and Qfreq are indicating "overrange" and DIN sensors are all showing open circuit as expected.

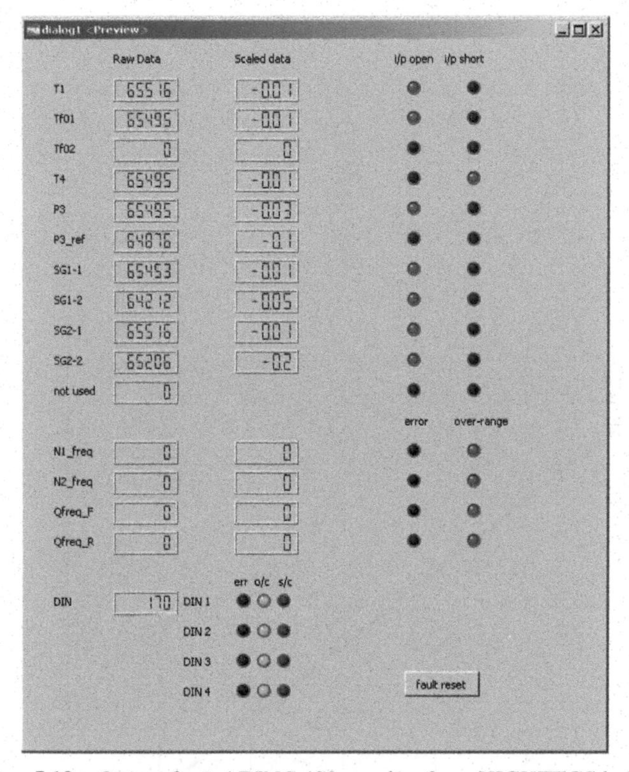

Figure 5.19 Output from ARINC 429 monitor from HIGHTECS hybrid.

Oscillations of the ARINC 429 signal were observed on most of the samples, which required additional capacitors on the input side of the supply regulators and the +ve and –ve power supplies needed to be applied simultaneously to avoid an overload effect on the hybrid circuit.

Tests have been carried out on the HIGHTECS hybrids to identify which connections are needed to provide an output on the ARINC 429 reader. The tests have shown that the following connections need to be applied to the HIGHTECS hybrid circuit:

- ADCSTARTX and ADCRESNX tied to Vdd
- ATEST, ADC_CLKX and ASEX0-ASELX3 tied to GND

The outputs from the TFo1 analog sensors on the ARINC429 reader are shown in Figure 5.20.

The results indicated that the analog sensors were working but the results were affected by the non-linearity in the ADC performance, which caused the spurious readings.

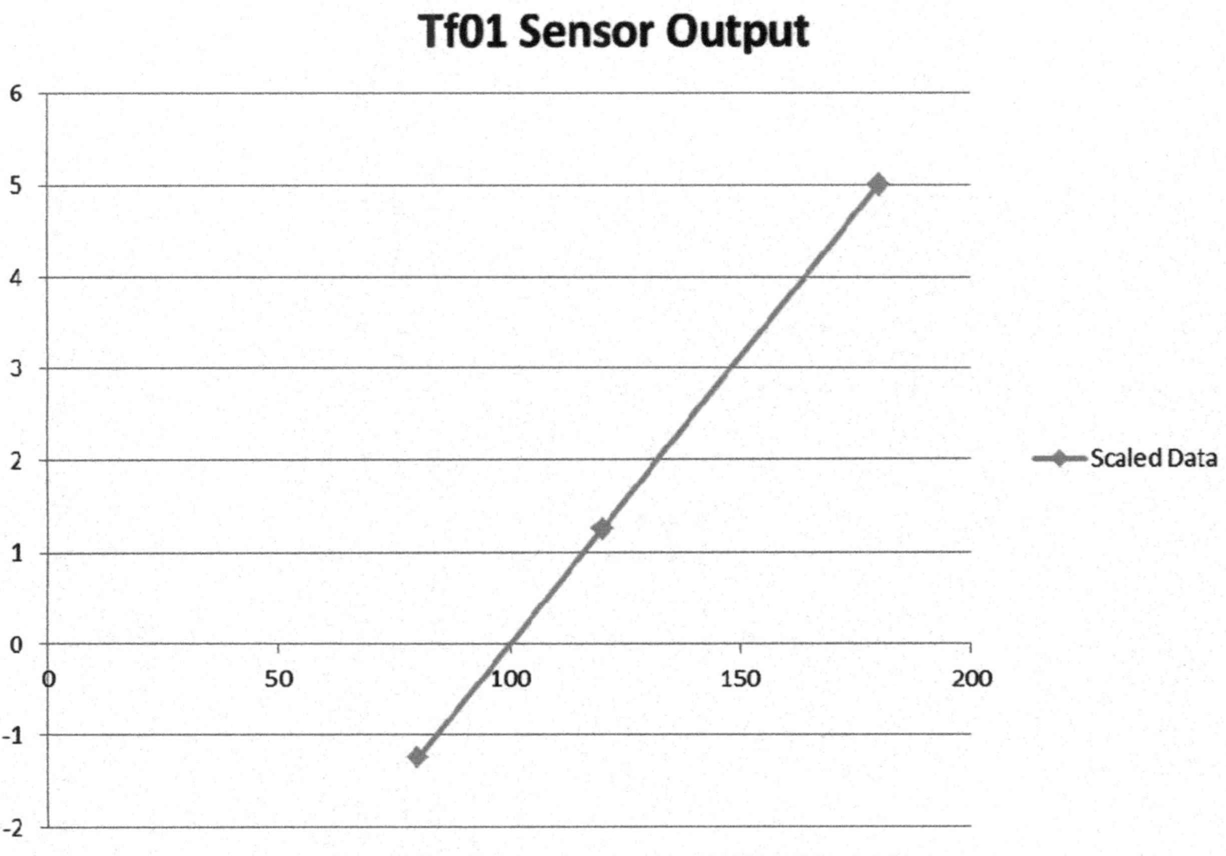

Figure 5.20 Tfo1 sensor output from HIGHTECS hybrid.

Testing of the Qfreq sensor was carried out using two pulse generators that were set up to provide two pulses, see Figure 5.21. The graph showing the response against frequency is shown in Figure 5.22, where the measured accuracy was better than 0.03% at room temperature.

The output from a representative HIGHTECS hybrid circuit with all sensor inputs open circuit connected to the ARINC 429 Bus Monitor User Interface is presented in Figure 5.19. The red LEDs are all indicating open circuit errors as expected. The Tfo2 signal has no error message, which was as expected as no ARINC messages were generated for this sensor. T4 signal is indicating an open circuit. Nfreq and Qfreq are indicating "overrange" and DIN sensors are all showing open circuit as expected.

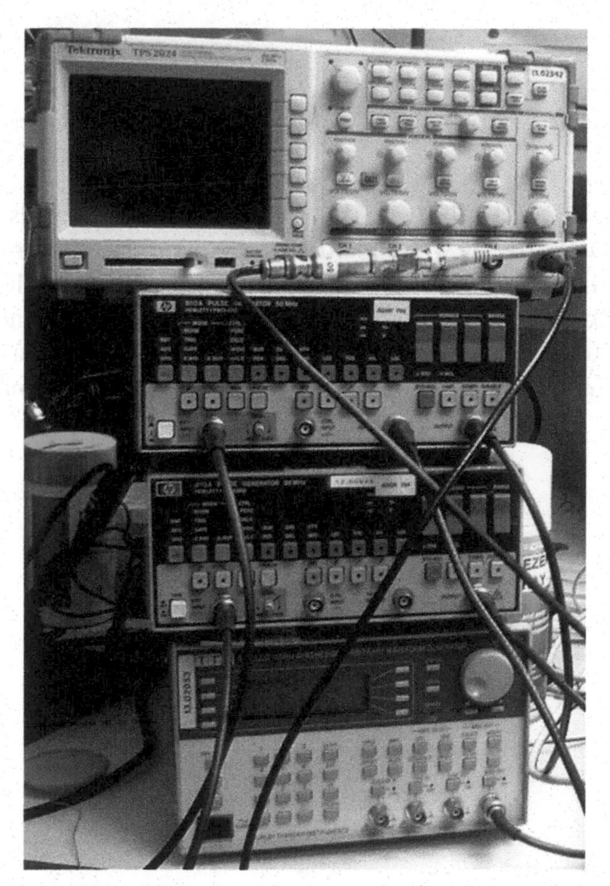

Figure 5.21 Pulse generators used for testing of Qfreq sensor.

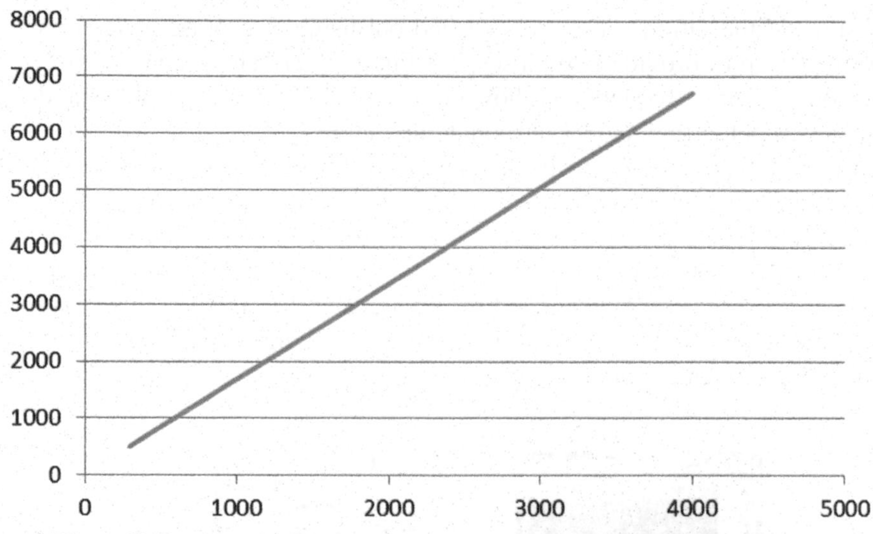

Figure 5.22 Qfreq sensor output against input frequency at room temperature.

Further work on the HIGHTECS hybrid to identify the signals required to operate the ADC correctly was undertaken, which produced near-expected outputs for the linear sensors and the frequency sensors. One HIGHTECS hybrid circuit (10514605) was assembled into a complete module and the unit has been shown to function from –40°C to +225°C, with the linearity of the SG2 sensor output improving as the temperature increases above ambient, see Figure 5.19.

6

Reliability, Failure Rates and Lifetime Prediction

6.1 Accelerated Life Tests and Lifetime Prediction

6.1.1 Thermal Ageing at 200°C and 250°C

Based on the temperature profiles supplied by Turbomeca, estimates of the operating lifetime expected by extrapolating results from temperature storage tests at 200°C and 250°C for 1000 hours have been made and are presented in Table 6.1.

6.2 FMEA and Reliability Prediction

A preliminary FMEA has been carried out based on the functional block description of the design of the HIGHTECS module containing the HIGHTECS hybrid circuit and a high temperature pcb containing resistors (2).

The main failure modes that may result in undetected erroneous data being sent are improper operation of the ADC on the ASIC, the voltage regulators and drift in capacitor and resistor values. The probability of erroneous data transmission is mainly controlled by the ability of the BIT function to flag warnings of when the various functional blocks do not function correctly.

The estimated values derived from the FMEA for the two temperature profiles provided by Turbomeca are presented in Table 6.2.

6.2.1 Module Weight and Dimensions

A breakdown of the weight of the prototype HIGHTECS Module is presented in Table 6.3.

The target weight for the HIGHTECS module was 500 gms, the actual weight was 986 gms of which over 80% was accounted for by the stainless

Table 6.1 Estimate of operating lifetime after extrapolation of temperature storage results for 1000 hours at 200°C and 250°C

Temperature Profile Supplied by Turbomeca	Average Operating Temperature	Storage Test Temperature	Test Time	Estimated Lifetime
1	84.4°C	200°C	1000 hours	61 years
		250°C	1000 hours	22 years
2	68.5°C	200°C	1000 hours	298 years
		250°C	1000 hours	109 years

Table 6.2 Summary of values derived from FMEA on HIGHTECS module

Factor	Temperature Profile 1	Temperature Profile 2
Total failure rate for HIGHTECS Module	$50.69/10^6$ flight hours	$41.42/10^6$ flight hours
Mean time between failures	19,730 hours	24,143 hours
Probability of no data transmitted	15.89×10^{-6} flight hours	12.98×10^{-6} flight hours
Probability of undetected incorrect data transmission	1.57×10^{-6} flight hours	1.29×10^{-6} flight hours
BIT failure detection cover	91.7%	91.6%

Table 6.3 Breakdown of weight by component for prototype HIGHTECS module

Component	Weight, gms
Stainless steel enclosure (exc connectors)	606
Stainless steel lid	113
Connectors	117
Mounting plate	48
High temperature pcb with resistors	48
Hybrid circuit (containing resistors)	38
Miscellaneous (washers, gaskets, etc)	16
Total	**986**

steel enclosure, lid and connectors. A significant reduction in weight of the HIGHTECS module can be achieved through selection of lighter materials (e.g. aluminium) for the enclosure and lid, although plating of the aluminium may be necessary to withstand the environment.

The target and actual dimensions of the prototype HIGHTECS module is presented in Table 6.4.

The actual dimensions of the prototype HIGHTECS module exceed the target dimensions, mainly on the length and width due to the currently available high temperature connectors. If miniature high temperature connectors are

Table 6.4 Target and actual dimensions for prototype HIGHTECS module

Dimension	Target mm	Actual mm
Length (inc connectors)	90	157.60
Width	40	64
Height	60	38.20

Table 6.5 Target and actual current power consumption for prototype HIGHTECS module

Consumption	Unit	Target	Actual
Power	W	10	2
Current	A	1	0.2

developed, there is scope for size reduction. Internally, the derated resistors for high temperature operation have the largest dimensions. As miniaturised high temperature resistors become more widely available, these resistors could be incorporated in the hybrid circuit.

6.2.2 Module Power Consumption

The target and actual power consumption of the prototype HIGHTECS module are presented in Table 6.5.

6.3 Summary

The HIGHTECS ASIC, hybrid circuit and module have been designed and manufactured. The HIGHTECS ASIC has successfully demonstrated dual output of ARINC 429 messages; however, problems have been encountered in achieving a consistent linear output in the Analogue to Digital Conversion (ADC) transfer function. The hybrid circuit and module has also produced ARINC 429 messages, but the output has been inconsistent, which again is believed to be related to the ADC transfer function. The ADC, which was supplied to the project as an existing IP block, is sensitive to its supply voltages and does not meet its published specification. The transfer function of the ADC has discontinuities present. The discontinuities reduce as the analogue supply voltage is increased above the digital supply voltage and as the temperature is increased above ambient. The voltages needed to eliminate the discontinuities are above those recommended for the SOI ASIC process. A small number of devices were identified which had a functioning ADC at a digital voltage of 5 V and analogue voltage of 5.5 V and these devices have been assembled into the HIGHTECS hybrid circuit and module. The results show that the

HIGHTECS module can function between −40°C and +225°C, with linearity of output improving as the temperature increases. A re-spin of the ASIC design was carried out to address the issues of the inconsistent ADC functionality by bringing out separate voltage references and improving the connections around and to the ADC block. The results on the 2nd version of the HIGHTECS ASIC show the analogue sensor conditioning and frequency measurements functions in line with specification on the ASIC over the temperature range −40°C up to 250°C with operation up to 275°C. However the ADC output is not linear at 5 V, which is the recommended voltage for the SOI process and further work will be required outside the scope of this project to develop an improved ADC IP block which can function at 5 V.

7

Future Directions for High Temperature Electronics

7.1 Semiconductor Devices

The preceding chapters in this book have demonstrated the capabilities for designing, manufacturing and testing an application specific semiconductor device based on SOI process technology for application in high temperature aero-engine control. The SOI semiconductor process suitable for high temperature operation is not widely available, with a limited number of foundries worldwide. As demand for niche applications with their low production quantities does not interest mainstream semiconductor companies, this situation is likely to remain in the future, unless even this demand is insufficient, causing existing foundry capabilities to be declared obsolete.

Alternative semiconductor materials and processes are being developed for high temperature applications, such as SiC and GaN, but the process complexity is restricted and further developments will be required to reach the current capability of the high temperature SOI process. As the markets will continue to be niche, low production quantities, the pace of device capability development will be governed by the ability to fund specialist application requirements rather than a widespread demand for the technology.

7.2 Passive Components

In parallel to the development of semiconductors for high temperature applications, advances have been made in extending the capabilities of passive components, such as capacitors, resistors and inductors.

For capacitors, ceramic based high dielectric materials have been developed with operating temperatures of up to 300–400°C, although derating of capacitors values need to be taken into account and lifetimes at these temperatures under voltage bias need to be established. Silicon capacitors are also attractive for stability at operating temperatures of up to 250–300°C. Overall capacitor values at temperatures of >250°C are limited to <10 μF and there does not seem to be many prospects for higher value capacitors operating above this temperature.

For resistors, precision thin film resistors are available for operation up to 250°C, with temperature limitations imposed by the materials used in the assembly of the resistors, including resistance shifts of the thin film resistor on ageing and deterioration of the high temperature adhesive used to attach the thin film resistor to a ceramic substrate. For general resistors, thick film resistor materials can be relatively stable up to 400°C. Other passive components, such as inductors, are available from a limited number of suppliers with an upper temperature limit of around 250°C.

7.3 1st and 2nd Level Assembly

For long-life products, such as required in aero-engine controls, the durability of the materials and connections used in the assembly of systems needs to be established, covering 1st and 2nd level processes. Long-term ageing tests have been carried out at 250°C, showing negligible deterioration for ageing periods of up to 1 year, but with predicted lifetimes of 25 years, higher temperature ageing studies to provide accelerated degradation factors are required. Within the demonstration unit, Al-1%Si wires were used to interconnect the ASIC to the metallisations on the package/substrate. Although the wire bonds were stable at 250°C, at temperature exposures of >300°C, the Al based wires would soften and alternative wires such as Au and Pd would need to be examined, along with custom compatible metallisations. For die attach and surface mount passive components, most high temperature adhesives do not provide long term durability at temperatures of 250°C and above. Attach with non-organic materials such as Au-Si eutectic, Ag-glass or possibly sintered Ag is recommended. For the 2nd level assembly processes, high melting point Pb based solders are used, although there is interest in developing Pb free alternatives, but there are few materials available capable of operating above 250°C.

7.4 Custom Metallisations

High temperature devices and components are normally supplied with a standard metallisation, based on the accepted practices of the manufacturer.

These metallisations are not always the most suitable for high temperature application when the interconnect materials are considered. As a rule of thumb, mono-metallic systems connecting the device/component to the package/substrate are desirable, but rarely achievable and a compromise has to be reached. In addition, diffusion from within the metallisation structure must also be taken into account for possible interaction between the connection and the device/component/substrate/package material, which can cause deterioration over time. It is recommended that device/component/substrate/package metallisations and interconnect systems are thoroughly reviewed to ensure compatibility or to conduct tests where there are doubts about long-term durability. It is possible sometimes to request custom metallisations from the device/component/package/substrate supplier, normally with a price penalty and subject to Minimum Order Quantities.

7.5 EMI/Lightning Protection

Alongside the specific devices and components for high temperature aero-engine control systems, protection against transients caused by EMI and lightning strikes must be catered for. At present, there is a dearth of components that can fulfil this function. Specialist devices (normally based on SiC) are in development at the major aero-engine manufacturers, although these devices are not yet proven and qualified. Until devices become available, EMI/Lightning protection will need to be provided away from the hot zone containing the engine control system, with additional costs, weight and losses of cabling.

7.6 Applications

The aero-engine control system developed and demonstrated has several common features that could be applied to other multi-sensing systems in different industry sectors, including down-well exploration and monitoring, gas turbine instrumentation, automotive engine and braking control systems and geothermal extraction. Although the environmental requirements for each application are different in detail, the overall design selection will always be based on integration of off-the-shelf components or custom design on silicon through an ASIC. The building blocks used in this ASIC can be reused in other applications, thus cutting down on design time and making the design process less application specific.

7.7 Commercial/Environmental Factors

7.7.1 Market Size

The market size for high temperature electronics based on aerospace and down-well applications has been growing gradually for many years. Although this growth in applications is positive, it is insufficient to attract significant interest from the major semiconductor foundries and supply of devices will remain in the realm from the niche semiconductor manufacturers at relatively high prices. An increase in demand for high temperature semiconductor devices from the automotive sector will broaden the range of foundry capabilities and reduce prices, but not to the extent of commodity prices seen in consumer electronics.

7.7.2 Custom vs Discrete Solutions

The product development cycle in most electronics applications normally involves breadboarding a particular solution using discrete off-the-shelf devices and components mounted onto a printed circuit board, before progressing towards an ASIC if the production quantities justify the design and manufacturing costs against a lower unit cost. In the field of high temperature electronics, the range of off-the-shelf devices is limited and the combination of discrete devices may not satisfy the application requirement. This situation leads towards adoption of custom electronics through an ASIC design earlier in the product development cycle. Design re-use or common building blocks which can easily incorporated into an ASIC will reduce the design time and costs, but there are no current examples of a high temperature gate array approach, where customisation takes place within the top metal layers during ASIC manufacture. Multi-Project Wafers are also not that common, as new high temperature designs are sporadic and the high value nature of projects in aerospace and down-well applications will normally justify a dedicated engineering wafer run, which, if successful, can also satisfy initial production quantities.

7.7.3 Integration into Systems

The connection of the high temperature electronics control unit into the overall system will normally by achieved through a robust connector (e.g. MIL-DTL-38999), some of which can operate up to 250°C, with the correct selection of materials. If higher temperature connections into the system are required, specialist lead/wire brazing/welding techniques may be required, which would need to be implemented on a case-by-case basis.

7.7.4 Lifetime Support

The lifetime of the product is dependent on the temperature profile experienced by the electronics control unit in service and any other environmental factors (e.g. vibration) that may accelerate deterioration of the unit. The desire is to have electronics that is based on a "fit and forget" principle, but the reality in high temperature electronics is that units may have to be replaced at some stage during the product lifetime. In aerospace, regular checks on the unit performance can be made during scheduled maintenance and replacements are possible. In down-well drilling applications, this can be achieved relatively easily between drilling operations, but for permanent monitoring operations, replacement will be difficult if not impossible.

7.7.5 Economics

The current status of the high temperature electronics market of high value, niche products means that price is not as key as experienced in commodity markets. The limited number of suppliers, low production quantities and specialist materials/processes leads to unit prices significantly higher than other industrial and consumer electronics. In some sectors, uprating of industrial electronics to operate beyond their specified upper temperature limit can satisfy the need for high temperature electronics with short lifetime requirements and provide a more cost effective solution.

In addition to the unit price considerations, the impact of using high temperature electronics can have an overall positive cost benefit for the system. For example, in aerospace, a reduction in weight can lead to fuel savings, and in down-well drilling, increasing the duration of a drilling operation can minimise downtime. Each case needs to be reviewed not just on the unit price, but on the overall system to assess whether there is an economic advantage in investing in a high temperature electronics solution.

Index

A

Analogue to Digital 13, 34, 85, 121
Application Specific Integrated Circuit 13
ARINC 22, 91, 97, 109

C

CMOS SOI 12, 13, 69, 70
Complementary Metal Oxide Semiconductor (CMOS) 11, 12, 69, 86
Current mirror 22, 41, 46, 71

E

Electrostatic Discharge 79
Engine Control Unit 8
Engine Health Monitoring System (EHMS) 8, 72

F

Full Authority Digital Engine (or Electronic) Control 8

H

High Temperature Co-Fired Ceramic 24, 86
High temperature electronics 3, 7, 19, 123

I

Instrumentation amplifier 31, 34, 55, 61

M

Metal Oxide Semiconductor Field Effect Transistor (MOSFET) 12

O

opamp 32, 34, 36, 38

P

Peak detector 38, 70, 71, 80
Pin Grid Array (PGA) 24, 95, 99, 111
Printed Circuit Board 24, 68, 95, 126

S

Silicon Carbide 12
Silicon on Insulator 20
Single-ended to differential converter 50, 53, 54, 56

About the Authors

Dr. Lucian Stoica (SM'14) received the B.Sc., M.S. degrees from Technical University of Iasi, Iasi, Romania, in 1999 and 2000, respectively.

He received the Dr. Tech. degree from University of Oulu, Oulu, Finland in 2008.

Since 2010, he is a Research Engineer and Project Leader with GE, Global Research, Munich, Germany.

His main research interests are robust integrated circuits and systems for harsh environment applications.

He has authored or co-authored over 15 publications in international journals or conferences, contributed to 1 book and 2 patents.

Steve Riches received a BA (Hons) in Natural Sciences from the University of Cambridge and has 16 years experience (1983–1999) in research and development at the Welding Institute (TWI) on interconnection and packaging of electronic devices and laser processing.

Since 1999, he has had 16 years experience in business development at GE Aviation Systems – Newmarket.

He has managed several UK DTI (TSB) collaborative projects and an EU Clean Sky project in last 3 years on high temperature electronics packaging involving several Universities, Institutes and Industrial Partners and presented regularly at high temperature electronics conferences in the USA and UK.

He left GE Aviation in December 2015 and is now a director at Tribus-D Ltd, a start up company.

Dr. Colin Johnston is Director of commercial and industrial services for the Department of Materials, University of Oxford and manager of Oxford Materials Characterisation Service. Colin holds a senior research fellowship in the Department of Materials, University of Oxford where he is active in research and development of novel materials solutions for high temperature and high reliability electronics packaging and energy storage materials. Colin is co-chair of the HITEN and HiTEC international conferences on high temperature electronics. He has over 100 peer reviewed publications with an RG score of 35.56 and h-index of 23.

Colin gained his Ph.D. in surface science and catalysis at Dundee University in 1987. He then went on to work for AEA Technology for 14 years working in diverse areas of materials research and development. He has over 15 years experience in technology translation and exploitation of university research most recently through his leadership role in the Materials Knowledge Transfer Network.

About the Contributors

Geoff Rickard graduated from Southampton University with a B.Sc. in Electronic Engineering and a M.Sc. in Computer Science.

He has more than thirty years of experience in IC design and is the author of 3 patents.

From 2006 to 2014, he was a Lead Design Engineer with GE Aviation, Cheltenham.

He retired in 2014.

Ozan Iskilibli has got his M.S. degrees in Communications Engineering from University of Technology, Munich, and Electrical & Computer Engineering from Carnegie Mellon University in 2012 and 2014 respectively. Having a strong foundation in solid-state electronics, he has worked on a variety of fields ranging from integrated circuit design to embedded programming and operating systems. He currently works for Oracle as a software engineer in Santa Clara, CA.

Paul Williams completed the Engineering Apprenticeship at Smith Industries between 1976–1980.

He joined Micro Circuit Engineering in 1980 as Layout Engineer and worked on full custom 3 um and 5 um designs.

From 1984 he worked as Design Engineer on approximately 100 ASIC and Gate Array designs in 5 um and 0.35 um processes.

Since 2002 he is responsible for all aspects of IC digital/analogue implementation and verification at GE Aviation, Cheltenham.

Reece Beigh received the Bachelor of Science in Electrical Engineering from the University of Washington in Seattle, Washington, USA in 2012, along with a Bachelor of Arts in Music.

He received his M.S. in Electrical and Computer Engineering from Georgia Institute of Technology in Atlanta, Georgia, USA in 2014, with a focus in control systems.

During his graduate studies he did an internship at GE Global Research in Munich, Germany.

In 2014, he joined MIT Lincoln Laboratory as engineering staff.

His current research interests include real-time operating systems and scheduling algorithms.

Renato Del Regno graduated with honors in Physics with a Master Degree specialization in Electronics.

His studies focused on digital electronics and FPGA programming.

During his Master, Renato worked as an intern student for GE Global Research in Munich, Germany where he experienced high temperature working of special ASIC circuits and VHDL programming of devices according to aeronautical data bus standards.

Afterwards, he worked in Italy as PCB layouter and in the field of calibration $\&$ testing of acquisition boards for avionics systems.

Dr. Thorsten Baumheinrich has 20 years of custom IC design experience in Si, SiGe, and III-V Technologies for optical communications, test & measurement and sensing applications.

He worked in Senior Engineering and technical Leadership roles at Texas Instruments, Inphi, Luxtera in the U.S. and at leading HighTech companies in Germany.

His research work is focused on speciality integrated circuit design for healthcare, security and sensing applications.

He was and is involved in EU-funded projects under the 4th and 7th framework programme and under the EU Horizon 2020 initiative.

Dr. Valentyn Solomko received the B.S. and M.S. degrees from the National Technical University of Ukraine and the Ph.D. degree (summa cum laude) from the Brandenburg University of Technology, Cottbus, Germany.

From 2009 until 2012 he worked at GE Global Research in Munich focusing on integrated and discrete electronic solutions for aviation, medical and industrial applications.

In 2012 he joined Infineon Technologies AG, Munich, developing integrated circuits and modules for RF front-ends in mobile handheld devices.

He is a visiting lecturer at the University of German Federal Armed Forces in Munich, lecturing a course on analog integrated circuit design.

Lightning Source UK Ltd.
Milton Keynes UK
UKOW06n1146270117
293027UK00002B/6/P

9 788793 379251